EVALUATION OF MINERAL RESERVES

APPLIED GEOSTATISTICS SERIES

General Editor
André G. Journel

André G. Journel and Phaedon C. Kyriakidis, *Evaluation of Mineral Reserves: A Simulation Approach*

Clayton V. Deutsch, *Geostatistical Reservoir Modeling*

Jean-Laurent Mallet, *Geomodeling*

Pierre Goovaerts, *Geostatistics for Natural Resources Evaluation*

Clayton V. Deutsch and André G. Journel, *GSLIB: Geostatistical Software Library and User's Guide*, Second Edition

Edward H. Isaak and R. Mohan Srivastava, *An Introduction to Applied Geostatistics*

Evaluation of Mineral Reserves
A Simulation Approach

André G. Journel
Phaedon C. Kyriakidis

OXFORD
UNIVERSITY PRESS

2004

OXFORD
UNIVERSITY PRESS

Oxford New York
Auckland Bangkok Buenos Aires Cape Town Chennai
Dar es Salaam Delhi Hong Kong Istanbul Karachi Kolkata
Kuala Lumpur Madrid Melbourne Mexico City Mumbai Nairobi
São Paulo Shanghai Taipei Tokyo Toronto

Published by Oxford University Press, Inc.
198 Madison Avenue, New York, New York 10016

www.oup.com

Library of Congress Cataloging-in-Publication Data
Journel, A. G.
Evaluation of mineral reserves : a simulation approach /
by André G. Journel and Phaedon C. Kyriakidis.
p. cm. — (Applied geostatistics series)
Includes bibliographical references and index.
ISBN 978-0-19-516694-1
1. Mines and mineral resources—Valuation—Statistical methods.
2. Mining geology—Statistical methods. I. Kyriakidis, Phaedon C.
II. Title. III. Series.
TN272.7.J68 2003
553—dc22 2003012482

Printed in the United States of America
on acid-free paper

To Harry Parker and Mohan Srivastava, two unsung giants.

Preface

This book was initiated by a survey that indicated the limitation of the 1990s literature with regard to published reports of applications of stochastic simulation in the mining industry. Many studies have used numerically simulated grade distributions to evaluate dilution factors, but almost none were published, because of lack of time or perhaps dedication. This statement may fire up those mining geostatisticians who are doing innovative work but did not care about publishing. They owe it to their ingenuity, to their clients if they are consultants, and to their students if they are teachers to document their thoughts and share their experience through publications with substance. Science progresses with wide open disclosure and debate, not with confidential notes or communications to forgettable forums, and certainly not with patented tradecrafts.

This book (re)introduces the concept of stochastic simulation as applied to evaluation of mineral reserves, more specifically evaluation of the impact of the support and information effects on mining dilution. The support effect relates to the difference in size between the volumes of the ore/waste units and the much smaller volumes of rock actually sampled that provide the sample data. The information effect relates to the difference in data quality and quantity between the time of reserves evaluation and that of grade control and actual mining. There are many other aspects of mine planning that affect the recoverable reserves, most prominently data accuracy and correct geological interpretation; these aspects are not covered in this book. Our goal is not to attempt a definitive essay on the science and art of mineral reserves evaluation, but to (re)introduce the stochastic simulation toolkit that provides a numerical framework for estimating many of the factors affecting mining dilution and reserves recovery.

Perhaps a good analogy to convey the potential of simulation is that of a wind tunnel where various designs of an aircraft wing are tested for resistance to turbulence. Our "wind tunnel" is a computer that stores various realizations of the spatial variability of mineral grades; our "aircraft design" is the suite of data acquisition, mine planning, and ore grade control processes that affect profitability; our "turbulence" is the set of unknowns: geological setting, grades, mining dilution, and so forth; and our "resistance" is the robustness of our planned mining operation in the face of all the above uncertainties. Stochastic simulation of mineral grades distribution followed by simulation of mining recovery processes has the potential of testing the adequacy of a mining project and tuning some of its parameters before committing a lot of time and money. The cost of conducting such a simulation is so minute compared to that involved in developing a new mining venture. We therefore submit that banks and regulatory agencies should require such an exercise before delivering any green light.

Chapter I introduces the aspects of mining dilution that will be addressed with the simulation approach developed in Chapter II. To provide a yardstick for comparison, the traditional volume-variance correction for difference of support volumes is recalled in Chapter III. Short of an exhaustive "real" mining data set, two reference data sets have been generated from actual quasi-exhaustive topographic data in Chapter IV, and the ore selection process has been mimicked on both of them. We have tried to remain as general as possible, documenting methodology rather than presenting real case studies; this explains our decision not to introduce explicit grade units. In Chapter V, the proposed simulation approach is applied to the previous reference data sets. The availability of an exhaustive reference, which is sparsely sampled to mimic actual mining data, allows comparison of predictions with actual reference values. In Chapter VI, the volume-variance correction algorithm is applied to the same sample data used in the simulation approach. In Chapter VII, some conclusions and recommendations are suggested.

All the data (reference and sampled) used in the case studies of this book can be downloaded over the Internet from the following Web site:

`http://ekofisk.stanford.edu/SCRFweb/mineral_reserves/index.html`

All kriging and stochastic simulation programs used can be found in the Geostatistical Software Library.

Finally, we would like to acknowledge the many friends who reviewed the numerous preliminary drafts of this book: Clayton Deutsch, Roussos Dimitrakopoulos, Jaime Gómez-Hernández, Pierre Goovaerts, Sia Khorowshahi, Ricardo Olea, Harry Parker, and Mohan Srivastava. This book could not have been written without their help; their numerous constructive criticisms indicate that the book shortcomings are ours only.

Contents

EVALUATION OF MINERAL RESERVES

Chapter I

Introduction

The problem of evaluation of future recoverable reserves is critical to the mining industry, in that reserves condition both investment and profitability associated with any mining venture. Recoverable reserves relate to a specific area/volume of the deposit, delineated by a mining plan to be economically accessible. This is as opposed to geological (in situ) resources, which refer to the amount of mineral available in the general area to be mined irrespective of whether this amount can be recovered (mined out) economically.

Statistical techniques for the evaluation of in situ resources on a panel-by-panel basis were proposed some 50 years ago by Krige (1951) and Sichel (1952). These pioneering works were later developed formally by Matheron (1962) into what is known as geostatistics, the application of which has now extended much beyond the mining arena. A few years later, Koch and Link (1970–71) and Davis (1973) published their seminal works on statistical analysis of geological data.

The difference between resources and reserves in geostatistical terms was clearly established by David (1977). A year later, Journel and Huijbregts (1978) proposed a formal approach to the problem of estimation of open pit reserves, addressing the respective influence of:

(i) The support of selection: ore is mined out as selective mining units of volumes vastly larger than the volumes of cores or composite grade data available at the time of the prediction study.

(ii) The information available at the time of selection, which is usually different from that available at the time of prediction.

This book addresses solely the two previous "predictable" dilution factors: support and information effect. It does not address unpredictable dilution and ore loss caused by numerous imponderables, such as geotechnical or hydrogeological conditions, the skill of miners in following diglines, political turmoil, and so forth. In any future reconciliation exercise, it is essential to distinguish between deviations from a mine plan resulting from predictable factors and those resulting from unpredictable factors.

A selective mining unit (SMU) is the smallest volume of ore that can be practically segregated from waste. The SMU is usually represented as a rectangular volume and, as such, is very tractable for numerical computations. In most deposits, however, ore selection is made using irregular areas/volumes that may be significantly larger than the SMU. The previous SMU definition implies that the volume recovered as ore will be an aggregate of SMUs; the definition of that volume cannot have resolution finer than the SMU size. In an open pit operation, the SMU could be the volume of influence of two to four blastholes.

One important aspect of reserves estimation, recognized but not addressed by the above early formalisms, was the "accessibility effect": a high-grade SMU may not be recoverable if the costs to reach it or to remove it selectively are prohibitive. In other words, there is rarely free selection of each SMU independent of the other SMUs which surround it. The amount of recoverable reserves depends on the mine plan (pit design, mine layout), which itself depends on the spatial distribution of mining unit grades.

The early solutions proposed by David (1977) and Journel and Huijbregts (1978) are still used for the evaluation of reserves some 25 years later. Except from a few notable works, for example, Journel (1974), Deraisme and Dumay (1979), Dagbert (1980), Kim et al. (1982), Chilès (1984), Deraisme et al. (1984), Journel and Isaaks (1984), Ravenscroft (1992), Nowak et al. (1993), Srivastava (1994a, 1994b), Guardiano et al. (1995), Caers et al. (1997), Dowd (1997), Glacken (1997), Nowak and Srivastava (1997), Dimitrakopoulos (1998), and Rossi (1999), the solution provided by conditional stochastic simulation is not widely used or is poorly published (as of 2003). In contrast, usage of conditional simulations in the petroleum industry and environmental sciences is better documented; see for example, Yarus and Chambers (1995), Baafi and Schofield (1997), Goovaerts (1997), Goméz-Hernández et al. (1999), Chilès and Delfiner (1999), Deutsch (2002), and the numerous references listed therein.

Mine planning is still typically performed on estimated maps of the grade spatial distribution; these maps provide a smooth and often biased image of the actual spatial distribution of the future SMU grades. Estimation of reserves is done either before mine planning, assuming free selection of each SMU, or after mine planning within the mine layout and again assuming free selection. A mine dilution/recovery factor is then applied to the estimated reserves. In fact, reserves evaluation and mine planning are intertwined issues, posing a difficult optimization problem (see Kennedy, 1990; Dowd, 1994). Last but not least, the traditional practice does not provide any formalism for assessing the uncertainty related to reserves estimation.

This book aims at presenting the conditional simulation toolkit, which allows generating alternative numerical models of the spatial distribution of grades, accounting for both the support effect and the information ef-

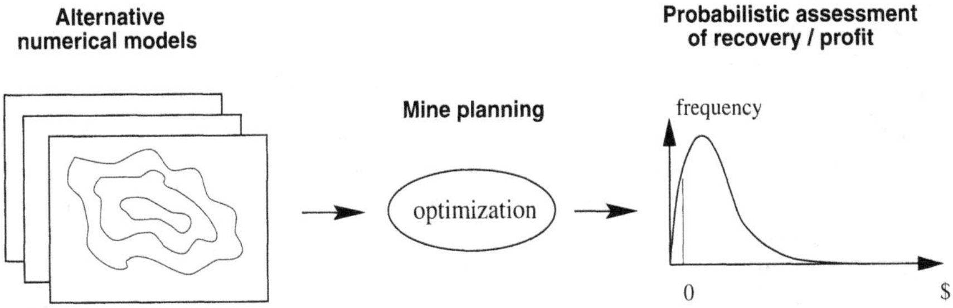

Figure I.1: Mining and recovery evaluation from alternative equiprobable numerical models of the grade(s) spatial distribution. An optimal mine plan and evaluation of corresponding recovery is applied to each numerical model, resulting in a probabilistic distribution of recovery (tonnage, metal, and profit).

fect. SMUs are selected on the basis of future grade estimates, the statistics of which are different from those of the presently available core grades. These numerical models are stochastic, in the sense that they are many, all "equiprobable", or more precisely, equally likely to be drawn. The difference between these alternative models reflects the uncertainty that is left once they are conditioned to the data available at the time of the prediction study. As opposed to maps of estimated values, conditional simulation provides images of the grade spatial distribution that are not smoothed. Such smoothing generally leads to underestimation of the proportion of SMUs with extreme high grades, and overestimation of the proportion of SMUs with low grades. Mine planning aimed at optimizing recoverable reserves and profit can be performed on each (or a selected few) of these alternative numerical models, resulting in not one but a distribution (histogram) of recoverable reserves and expected profit values, thus providing a measure of uncertainty (see Fig. I.1).

The difficult optimization problem underlying the joint definition of an optimal mine layout and the evaluation of the corresponding reserves is not addressed in this book. This problem is not likely to have a general solution: the solution will depend on the type of mine (e.g., underground vs. open pit) and specific technological constraints (e.g., rock mechanics, ventilation, dewatering, stockpiling). In all cases, however, conditional simulations provide a better basis (more realistic numerical models) to perform such an optimization (see Deutsch et al., 2000). For a perspective of the general problem of mine planning and reserves evaluation, readers may refer to King et al. (1982), David (1988), Kennedy (1990), Ravenscroft (1993), and Farrelly and Dimitrakopoulos (2002).

This book addresses the limited problem of evaluation of reserves within a given panel, under the assumption of free selection. This will be done using conditionally simulated models of the grade spatial distribution, and the results will be compared to predictions obtained using routine present-day geostatistical methodology.

The limitation of free selection within a given panel is necessary because, again, evaluation of reserves under technical constraints is an optimization problem specific to each set of such constraints. On a limited number of such panels, recovery under free selection can be simulated; in addition, the lesser recovery under specific technical constraints can be evaluated (see Norrena and Deutsch, 2001). The ratio of these two recovery figures would provide an assessment of these predictable aspects of mining dilution, to be applied to all other recovery figures obtained under free selection.

Section I.5 at the end of this chapter indicates the scope of this book and the spectrum of readers who may benefit from it.

I.1 Statement of the Problem and Notation

Selective mining operates on blocks or selective mining units (SMUs). The domain informed by each SMU; that is, its support, is denoted as v. In an open pit situation, the size of each SMU, denoted as $|v|$, can range anywhere from 3 m $\times$ 3 m to 20 m $\times$ 20 m $\times$ bench height. In an underground operation, the SMU size $|v|$ can be highly variable, ranging from a massive column in a block caving operation down to the volume of one scoop of a front-end loader in a highly selective operation. In the following, we will consider a constant SMU v, assuming that there is no resolution below that support v: each SMU or local aggregate of SMUs is sent entirely to the mill (it is recovered) or to the waste pile (its metal content is lost). One can rightfully argue that the concept of a single SMU size does not apply to an underground operation, in which case the examples given in this book only relate to open pit operations.

Recoverable reserves relate to the proportion (tonnage) and average grade of those SMUs that will be selected as ore within any given panel V of size $|V|$ (see Fig. I.2A). Prior mine planning would have determined which panel V would be subjected to selective mining. Free selection of SMUs of constant size $|v|$ is assumed within each panel V. However, the panel V could be of any size, at the limit as large as the whole deposit D or as small as to contain only one single SMU; the SMU size may vary from one panel to another.

The panel V should not be so large as to include widely different mineralization zones, each having different statistics (grade continuity and histogram). In geostatistical terms, a decision of stationarity should be reasonable within each panel. The panel size and geometry can vary depending on the extent of each mineralization zone considered. The panel V should not be so small as to include too few SMUs, rendering the histogram of SMU grades within V unreliable. Beware also that the smaller the panel V, the less the error averaging and the larger the uncertainty about the estimate of its recoverable reserves. In most applications, the whole panel V would be mined out, but only part of it will be sent to the mill or to the ore pile.

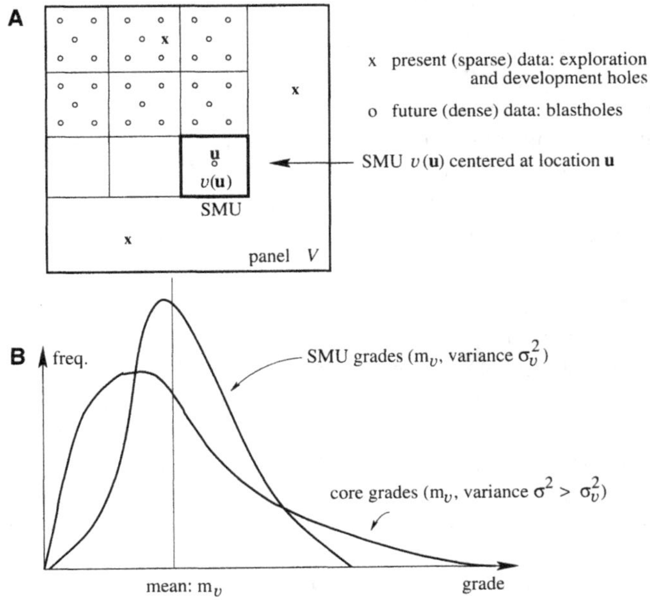

Figure I.2: The support and information effects. **A:** within a large panel V, selection is performed on estimated SMU grades of support volume $|v|$ much greater than the elementary core data volume. The selection estimate is based on future data available at the time of mining. **B:** the histogram of SMU grades (whether estimated or true values) is different from the histogram of sample grades.

Evaluation of recoverable reserves under free selection faces two major problems associated with:

1. **The support effect:** SMU grades relate to a volume of rock, orders of magnitude larger than the support volume of the core or composite grade data available at the time of the prediction study. In particular, the histogram of SMU grades has smaller variance and a different shape than the histogram of sample support grades within the same panel V (see Fig. I.2B). However, because any SMU grade is the linear average of its core grades (under exhaustive sampling), the two histograms will have the same mean, m_v.

2. **The information effect:** Actual selection is not performed on the true SMU grade but on some estimate of it; that estimate is typically based on the latest data available at the time of selection, including possibly a dense grid of grade control data; for example, blasthole samples in an open pit situation. That estimate is *not* the estimated SMU grade available at the time of prediction, when only exploration and development drilling is available.

Although any particular SMU is selected on the basis of its future estimated grade, it is the corresponding true grade (not its estimate) that is sent to the mill. The challenge of recoverable reserves estimation is to predict the number (tonnage) and average grade of those SMUs that will be recovered in the future at the time of mining, accounting for both support and information effects. Once again, free selection of each SMU within the panel V is assumed here; that is, each SMU is selected on the basis of its (future) estimated grade independent of any other SMU or any accessibility consideration within that panel.

We are aware that our definition of recoverable reserves departs somewhat from presently (2004) accepted definitions, such as those proposed by the Australasian Institute of Mining and Metallurgy (Joint Ore Reserves Committee [JORC] 1999). Per JORC, reserves imply recoverable; the only modifiers permitted reflect the degree of confidence— that is, proven or probable. The JORC definition relates to global reserves of either the entire mine or a large section of it. Our definition relates to local reserves over a much smaller area or panel V, and we do need the modifier between ore and waste, recoverable and noneconomical reserves, within that panel.

Although our proposed simulation approach does provide a numerical assessment of uncertainty about recoverable reserves (within a panel V), we did not attempt to extend that assessment into a classification of proven versus probable reserves. There are not yet any general geostatistical guidelines for such a classification at the local (panel) level.

Preliminary Notation (see also Appendix B).

• D: deposit, i.e., an assemblage of large panels V not necessarily all of the same size.

• V: panel, i.e., an assemblage of N selective mining units (SMUs) of equal size; from one panel to another the SMU size may vary. The size of panel V is denoted as $|V|$.

• v: SMU of constant size $|v|$ within any single panel V.

• $\mathbf{u} = (x, y)$: coordinate vector denoting a two-dimensional location within the deposit D. A a three-dimensional location would be denoted as $\mathbf{u} = (x, y, e)$, where e denotes elevation.

• $z(\mathbf{u})$: sample support grade at location $\mathbf{u}$: the core or composite support volume is assumed constant and quasi-point.

• $z(\mathbf{u}_\alpha)$: grade value at sample location $\mathbf{u}_\alpha$.

• n: total number of presently available grade samples for prediction of recoverable reserves. (n) denotes the set of samples.

• $n(\mathbf{u})$: number of presently available grade samples within a neighborhood around location $\mathbf{u}$. Note that $n(\mathbf{u}) < n$.

• nn: number of grade samples available at the future time of actual selection. Note that $nn(\mathbf{u}) > n(\mathbf{u})$.

• $z_v(\mathbf{u}) = \frac{1}{|v|}\int_{v(\mathbf{u})} z(\mathbf{u}')d\mathbf{u}' \simeq \frac{1}{n_v}\sum_{j=1}^{n_v} z(\mathbf{u}_j)$: true average grade of the SMU v centered at $\mathbf{u}$, with n_v being the number of points $\mathbf{u}_j$ discretizing the support v.

• $z_V(\mathbf{u}) = \frac{1}{|V|}\int_{V(\mathbf{u})} z(\mathbf{u}')d\mathbf{u}' = \frac{1}{N}\sum_{j=1}^{N} z_v(\mathbf{u}_j)$: true average grade of the panel V comprised of N SMUs of size $|v|$.

• $z_v^*(\mathbf{u})$: estimated SMU grade using the $n(\mathbf{u})$ data available at the time of prediction.

• $z_v^{**}(\mathbf{u})$: future estimated SMU grade using the $nn(\mathbf{u})$ data available at the time of selection. The three values $z_v(\mathbf{u})$, $z_v^*(\mathbf{u})$, and $z_v^{**}(\mathbf{u})$ are usually different.

• $z^{(l)}(\mathbf{u})$: simulated quasi-point support grade value at location $\mathbf{u}$. The superscript (l) stands for the lth simulated realization out of a total of L: $l = 1, \ldots, L$.

• $z_v^{(l)}(\mathbf{u}) = \frac{1}{|v|}\int_{v(\mathbf{u})} z^{(l)}(\mathbf{u}')d\mathbf{u}' \simeq \frac{1}{n_v}\sum_{j=1}^{n_v} z^{(l)}(\mathbf{u}_j)$: simulated true average grade of the SMU v centered at $\mathbf{u}$.

• $z_v^{**(l)}(\mathbf{u})$: simulated estimated SMU grade using the $nn(\mathbf{u})$ simulated data $\{z^{(l)}(\mathbf{u'}_\alpha),\ \alpha = 1, \ldots, nn(\mathbf{u})\}$, mimicking the information available at the time of selection.

> Knowledge of the pair of values $\{z_v(\mathbf{u}), z_v^{**}(\mathbf{u})\}$, $\forall v \in V$, would solve fully the problem, as one would know the actual average grade $z_v(\mathbf{u})$ of any SMU v selected on the basis of its future estimated value $z_v^{**}(\mathbf{u})$. At the time of prediction, however, both $z_v(\mathbf{u})$ and $z_v^{**}(\mathbf{u})$ are unknown. The known simulated pairs $\{z_v^{(l)}(\mathbf{u}), z_v^{**(l)}(\mathbf{u})\}$, $l = 1, \ldots, L$, provide a suitable proxy.
>
> Conditioning the simulated grade realizations ensures that the present data are reproduced:
>
> $$z^{(l)}(\mathbf{u}_\alpha) = z(\mathbf{u}_\alpha), \quad \forall l, \alpha = 1, \ldots, n$$

• z_c, or simply z: cutoff grade above which a SMU is declared as ore, i.e., sent to the mill.

> Perfect selection corresponds to the cutoff grade applied to the actual true SMU grade $z_v(\mathbf{u})$.

Actual selection corresponds to the cutoff grade applied to the estimate $z_v^{**}(\mathbf{u})$ available at the time of selection.

- $i_v^0(\mathbf{u}; z_c) = \begin{cases} 1 & \text{if } z_v(\mathbf{u}) > z_c \\ 0 & \text{if not} \end{cases}$ indicator of perfect selection

- $i_v^{**}(\mathbf{u}; z_c) = \begin{cases} 1 & \text{if } z_v^{**}(\mathbf{u}) > z_c \\ 0 & \text{if not} \end{cases}$ indicator of actual selection

- $t_v^0(z_c)$, $q_v^0(z_c)$, $m_v^0(z_c)$, $p_v^0(z_c)$: ore tonnage, quantity of metal, mean ore grade, and profit under perfect selection as functions of the cutoff grade z_c.

- $t_v(z_c)$, $q_v(z_c)$, $m_v(z_c)$, $p_v(z_c)$: recovery functions under actual selection performed on the estimates $z_v^{**}(\mathbf{u})$ available at the time of mining.

I.2 Perfect Selection

Ideal recovery would correspond to a perfect, faultless, selection based on the true SMU grades $z_v(\mathbf{u})$; this is *never* achievable in practice.

Let $i_v^0(\mathbf{u}; z_c)$ denote the indicator associated with perfect selection of a SMU v centered at $\mathbf{u}$, defined as:

$$i_v^0(\mathbf{u}; z_c) = \begin{cases} 1 & \text{if } z_v(\mathbf{u}) > z_c \\ 0 & \text{if not} \end{cases} \tag{I.1}$$

where z_c denotes the cutoff grade determined from economic considerations, and the superscript 0 indicates perfect selection obtained by applying the cutoff z_c to the true SMU grade $z_v(\mathbf{u})$.

Ore tonnage: The corresponding tonnage is defined as the number of profitable SMUs within panel V, recovered under perfect selection:

$$t_v^0(z_c) = \sum_{j=1}^{N} i_v^0(\mathbf{u}_j; z_c) \quad \in [0, N]\ , \ \mathbf{u}_j \in V \tag{I.2}$$

where N is the number of SMUs in the panel, with $|V| = N \times |v|$. The tonnage of one SMU is assumed unit (tonnage factor equal to 1).

Quantity of metal: Because the SMU tonnage is unity, the corresponding quantity of metal is defined as the sum of profitable SMU grades, again recovered under perfect selection:

$$q_v^0(z_c) = \sum_{j=1}^{N} i_v^0(\mathbf{u}_j; z_c) \cdot z_v(\mathbf{u}_j) \tag{I.3}$$

Mean ore grade: The corresponding mean ore grade is defined as the ratio:

$$m_v^0(z_c) = \frac{q_v^0(z_c)}{t_v^0(z_c)} \tag{I.4}$$

This is the truncated mean above the cutoff grade z_c of the histogram of the N true SMU grades $z_v(\mathbf{u})$ to be found in panel V.

Profit: Profit is a function of ore tonnage and grade, typically determined by various technical and economical parameters. In this book, the following simplified formula is adopted:

$$p_v^0(z_c) = C(z_c) \cdot \left[q_v^0(z_c) - t_v^0(z_c) \cdot z_c\right] - C_0 \tag{I.5}$$

where $C(z_c)$ is a proportionality factor related to the revenues and costs of moving and processing the ore tonnage, and C_0 is an initial capital investment. The values of the parameters $C(z_c)$ and C_0 are case-specific and are given later with the reference data set in Chapter IV. A better profit function would consider different proportionality factors for quantity of metal and ore tonnage.

The profit should be an increasing function of the quantity of metal processed $q_v^0(z_c)$ and a decreasing function of the ore tonnage $t_v^0(z_c)$. As long as these two order relations are verified, any other reasonable profit function could have been considered (Lane, 1988). Note that any misclassification (imperfect selection) would necessarily decrease the profit, and hence $p_v^0(z_c)$ represents an upper bound for any actual profit, although it could be less than the estimated profit (overestimation).

I.3 Actual Selection

In mining practice, at the time of selection each SMU $v(\mathbf{u})$ is declared as ore or waste on the basis of its ultimate estimated value $z_v^{**}(\mathbf{u})$, using all the $nn(\mathbf{u})$ data available in the neighborhood of $\mathbf{u}$ at that time. Because $z_v^{**}(\mathbf{u}) \neq z_v(\mathbf{u})$, actual selection is necessarily imperfect: some poor SMUs such as $\{z_v(\mathbf{u}) \leq z_c | z_v^{**}(\mathbf{u}) > z_c\}$ are incorrectly sent to the mill, and conversely, some rich SMUs such as $\{z_v(\mathbf{u}) > z_c | z_v^{**}(\mathbf{u}) \leq z_c\}$ are sent to the waste; see Figure I.3 for a graphical representation of that misclassification. The notation $\{z_v(\mathbf{u}) \leq z_c | z_v^{**}(\mathbf{u}) > z_c\}$ should be read as: The true grade $z_v(\mathbf{u})$ is no greater than the cutoff grade z_c, given that its future estimate $z_v^{**}(\mathbf{u})$ is greater than z_c.

Let $i_v^{**}(\mathbf{u}; z_c)$ denote the indicator associated with a SMU declared as profitable on the basis of the estimate $z_v^{**}(\mathbf{u})$:

$$i_v^{**}(\mathbf{u}; z_c) = \begin{cases} 1 & \text{if } z_v^{**}(\mathbf{u}) > z_c \\ 0 & \text{if not} \end{cases} \tag{I.6}$$

In an open pit operation, typically, the selection estimate $z_v^{**}(\mathbf{u})$ would be determined from blasthole or other grade control data available just before mining (see Fig. I.2A). In the case of blastholes, such data are much more numerous but often of poorer quality than the drillhole data available at the time of the prediction study.

In practice, the final delineation of ore in a mining operation calls for other intangible information than grade control data, such as visual inspection of the mining faces, experience of the geologist, and so forth. Such information, although undoubtedly critical, cannot be easily modeled mathematically, and hence will be ignored in this book.

Similar to the case of perfect selection (Section I.2), recovery functions corresponding to actual selection based on the estimates $z_v^{**}(\mathbf{u})$ are defined (the star superscript [single * or double **] generally connotes an estimated value; the double star superscript [**] relates to estimates built from future data available at the time of selection):

- **ore tonnage:**

$$t_v(z_c) = \sum_{j=1}^{N} i_v^{**}(\mathbf{u}_j; z_c) \quad \in [0, N] \tag{I.7}$$

- **quantity of metal:**

$$q_v(z_c) = \sum_{j=1}^{N} i_v^{**}(\mathbf{u}_j; z_c) \cdot z_v(\mathbf{u}_j) \tag{I.8}$$

[note that even though the selection indicator $i_v^{**}(\mathbf{u}; z_c)$ is based on the estimated value $z_v^{**}(\mathbf{u})$, it is the corresponding true SMU grade $z_v(\mathbf{u}_j)$ that is recovered]

- **mean ore grade:**

$$m_v(z_c) = \frac{q_v(z_c)}{t_v(z_c)} \tag{I.9}$$

- **profit:**

the actual profit is defined from the actual grades sent to the mill:

$$p_v(z_c) = C(z_c) \cdot [q_v(z_c) - t_v(z_c) \cdot z_c] - C_0 \tag{I.10}$$

A graphical representation

Figure I.3 gives a graphical representation of the joint distribution (scattergram) of the N pairs of values $\{z_v(\mathbf{u}_j), z_v^{**}(\mathbf{u}_j)\}$, $j = 1, \ldots, N$, corresponding to a given panel V. The corresponding marginal histograms are given to the left of the ordinate axis for $z_v(\mathbf{u})$ and below the abscissa axis for $z_v^{**}(\mathbf{u})$.

The following observations can be made:

- because of data sparsity, the distribution of the estimates z_v^{**} is usually smooth; that is, it has a smaller variance than the distribution of the true grades z_v.

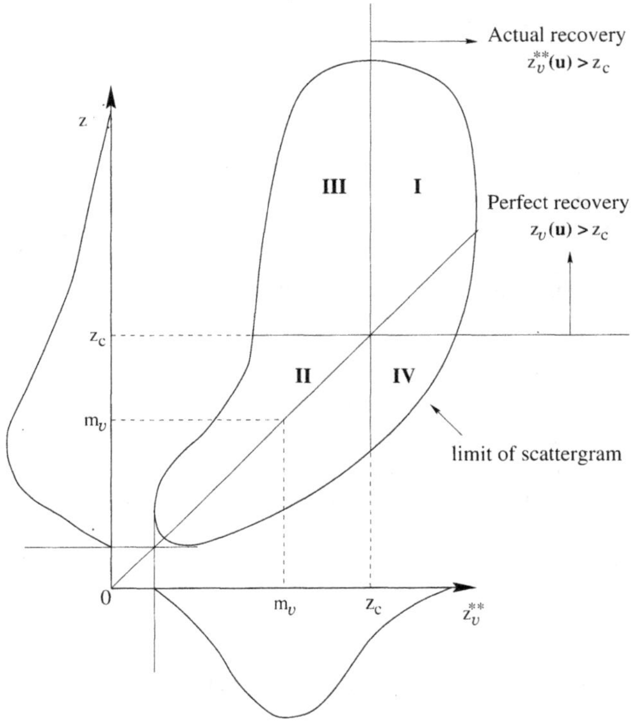

Figure I.3: Joint distribution (scattergram) of the true SMU grades $z_v(\mathbf{u})$ versus selection estimates $z_v^{**}(\mathbf{u})$. The two marginal distributions (histograms) are given at the left and below the axes. Note the smaller variance of the distribution of z_v^{**}. The cutoff value z_c defines four sectors on the scattergram: **I:** $\{z_v(\mathbf{u}) > z_c | z_v^{**}(\mathbf{u}) > z_c\}$: SMUs correctly classified as ore, **II:** $\{z_v(\mathbf{u}) \leq z_c | z_v^{**}(\mathbf{u}) \leq z_c\}$: SMUs correctly classified as waste, **III:** $\{z_v(\mathbf{u}) > z_c | z_v^{**}(\mathbf{u}) \leq z_c\}$: SMUs misclassified as waste, **IV:** $\{z_v(\mathbf{u}) \leq z_c | z_v^{**}(\mathbf{u}) > z_c\}$: SMUs misclassified as ore.

- the histograms and scattergram are typically positively skewed with more values below the common mean m_V. The histogram of z_v^{**} has a shorter tail than that of z_v: this is the smoothing of extreme grades z_v caused by estimation.

- the area to the right of the vertical $z_v^{**} = z_c$ corresponds to the SMUs actually recovered with estimated grades $z_v^{**}(\mathbf{u}) > z_c$. That actual recovery area is split into two sectors I and IV: the first (I) corresponds to SMUs correctly classified as ore, whereas the second (IV) corresponds to SMUs misclassified as ore.

- an efficient selection at the cutoff z_c is one that minimizes the misclassification sectors III and IV of the scattergram. Perfect selection corresponds to $z_v^{**} = z_v$: the scattergram then collapses onto the 45° line.

- the distribution of z_v values (ordinate axis) depends on the support of the SMU, whereas the distribution of z_v^{**} values (abscissa axis) depends essentially on the amount and precision of the future data, as well as on the estimation algorithm used to derive the estimates $z_v^{**}(\mathbf{u})$. Any change of the support size $|v|$ or the level of information available for estimation would result in a new scattergram $\{z_v, z_v^{**}\}$. This scattergram represents the interaction of support and information effects on the estimation of recoverable reserves.

In practice, the scattergram of Figure I.3 cannot be observed (unless postproduction), because the true grades z_v and the ultimate selection estimates z_v^{**} are not available at the time of the prediction study. Proxies of that scattergram, however, can be built from pairs of simulated values $\{z_v^{(l)}(\mathbf{u}_j), z_v^{**(l)}(\mathbf{u}_j)\}$, $j = 1, \ldots, N$. There can be L such proxies if L simulated realizations are generated. This is the very foundation of the proposed approach for evaluating recoverable reserves, as developed later in Chapters II and V.

I.4 Estimation and Simulation

Before presenting the various approaches to estimation of recoverable reserves, it is worth recalling the difference between an estimated field, say, the distribution in space of estimated SMU grades $\{z_v^*(\mathbf{u}), \mathbf{u} \in D\}$, and a simulated field $\{z_v^{(l)}(\mathbf{u}), \mathbf{u} \in D\}$. Both fields aim at representing the spatial distribution of true SMU grades $\{z_v(\mathbf{u}), \mathbf{u} \in D\}$, and both are built from the same data, say the n sample data available at the time of prediction (see Figs. I.4–I.6).

The expert reader may skip this section or refer to the excellent discussions on the same topic by Ravenscroft (1992) and Goovaerts (1997, chap. 8). Short of exhaustive information, in which case trivially all estimates or simulated values identify the corresponding true values, any attempt at representing an unknown true value $z_v(\mathbf{u})$ must adopt a specific goodness criterion: an estimate that is good for one criterion achieves this status at the expense of being poorer for a different criterion.

Typically, estimation (e.g., kriging) aims at local accuracy, whereas simulation aims in addition at structural accuracy. Each single estimated value, say $z_v^*(\mathbf{u})$, aims at being closest to the target unknown value $z_v(\mathbf{u})$. Simulation, instead, forfeits some local accuracy for providing a simulated field $\{z_v^{(l)}(\mathbf{u}), \mathbf{u} \in D\}$ that mimics structural characteristics of the true field $\{z_v(\mathbf{u}), \mathbf{u} \in D\}$. For example, simulation aims at reproducing the histogram and variogram model inferred from the sample data. Postmortem, the scattergram of the pairs of estimated-true values, $\{z_v^*(\mathbf{u}), z_v(\mathbf{u})\}$, would display better correlation than the corresponding scattergram of simulated-true values

$\{z_v^{(l)}(\mathbf{u}), z_v(\mathbf{u})\}$. However, the histogram and variogram of the simulated values $\{z_v^{(l)}(\mathbf{u}), \mathbf{u} \in D\}$ would be closer to the histogram and variogram of the true values $\{z_v(\mathbf{u}), \mathbf{u} \in D\}$. A simulated field is more likely to display the spatial variability, and hence the structures, of the true values. The question is: Which type of accuracy, local or structural, is more relevant to the evaluation of recoverable reserves?

For selection at the ultimate time of mining, local accuracy is required. For ore grade control one needs to minimize misclassification when selecting any particular SMU $v(\mathbf{u})$. However, at the much earlier time of a feasibility assessment and prediction of global recoverable reserves, there is no need for, nor should there be any expectation for, accurately locating every single ore SMU. Instead, it is important that the histogram of simulated values $\{z_v^{(l)}(\mathbf{u}), \mathbf{u} \in D\}$ displays the same proportion of high-grade SMUs (ore) as the histogram of true SMU grades $\{z_v(\mathbf{u}), \mathbf{u} \in D\}$. Similarly, it is important that the spatial variability of the simulated representation mimics that of the true SMU grades. The worst case would be an estimated representation that overstates the actual spatial continuity of good ore zones, because such oversmooth representation would lead to predicting better selectivity and recovery than can be achieved in the future.

Figure I.4A gives the grayscale map of a synthetically generated field of 50×50 values, which could represent the spatial distribution over a bench (or panel V) of $N = 2500$ "true" SMU grades $z_v(\mathbf{u})$, say gold grades expressed in grams per tonne. The size of each SMU is 4×4 meters. Figures I.4B and I.4C give the histogram and the standardized omnidirectional semivariogram (i.e., averaged over all horizontal directions and divided by the variance of the $N = 2500$ values) of these true SMU grades. Note the mean value $m_V = 2.58$ and the proportion $p(> z_c) = \frac{1}{N} t_v^0(z_c) = 0.25$ of SMUs with true grades above the cutoff $z_c = 2.56$; refer to the definition of ore tonnage under perfect selection (Equation I.2).

Figure I.5A gives the spatial posting of a sparse sample of $n = 10$ quasi-point support data; that is, data whose support volume is much smaller than that of the SMU grades mapped in Figure I.4A. Using that sample data set, estimation (here block kriging) was performed resulting in the estimated representation $\{z_v^*(\mathbf{u}), \mathbf{u} \in V\}$ shown in Figure I.5B. That representation is as locally accurate as the sparse sample data permit (see the scattergram of true versus estimated values $\{z_v(\mathbf{u}), z_v^*(\mathbf{u})\}$ displayed in Fig. I.5D). The histogram of the $N = 2500$ estimated values, shown in Figure I.5C, reveals a severe smoothing effect: the variance is only $(0.63)^2$, compared to the variance $(5.15)^2$ of the corresponding 2500 true values (from Fig. I.4B). More importantly, the proportion of estimated values $z_v^*(\mathbf{u}) > z_c = 2.56$ is 29%, compared to the 25% of such true values. This bias (called conditional bias) could affect severely any estimation of recoverable reserves over the entire panel V: the tonnage is overestimated by 16%, and the recovered grade is underestimated by 59%.

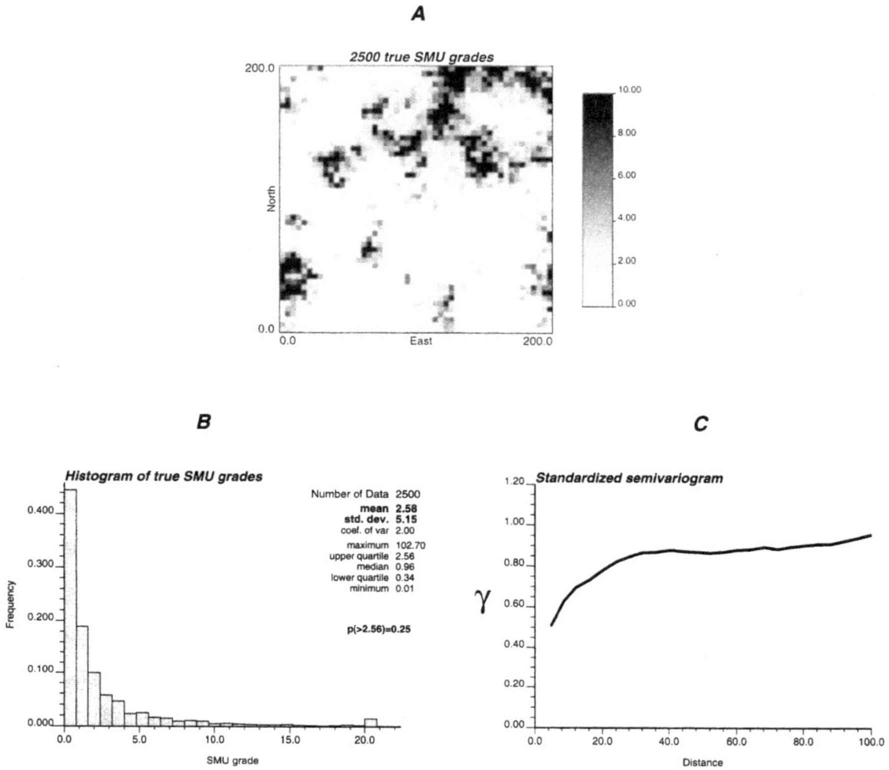

Figure I.4: Map of reference "true" SMU grades (**A**), their histogram (**B**), and omnidirectional semivariogram standardized to a unit sill (**C**), all taken from the geostatistical software library (GSLIB) data set (Deutsch and Journel, 1998).

Where the few sample data show high grades, the estimated SMU grades would display uniform patches of high values that do not reflect the actual spatial variability of the true SMU grades: the map of estimated values (Fig. I.5B) is much smoother than the true map of Figure I.4A. This smoothing effect is reflected by the variogram of estimated values (see Fig. I.5E). The variogram of estimated grades at short lag distances is much lower than the actual true variogram, indicating that the estimated grades display a much more continuous spatial distribution than the true grades. Both variograms in Figure I.5E are standardized to their respective variances; the nonstandardized variogram of estimated values would display much lower variance at all separation vectors $\mathbf{h}$ than the variogram of true grades.

What is needed is a SMU grade representation that would reproduce the histogram and semivariogram model adopted for the true SMU grades (Figs. I.4B and I.4C). Such statistics can be obtained from the geostatistical analysis of sample data, even if these data are defined on a support volume

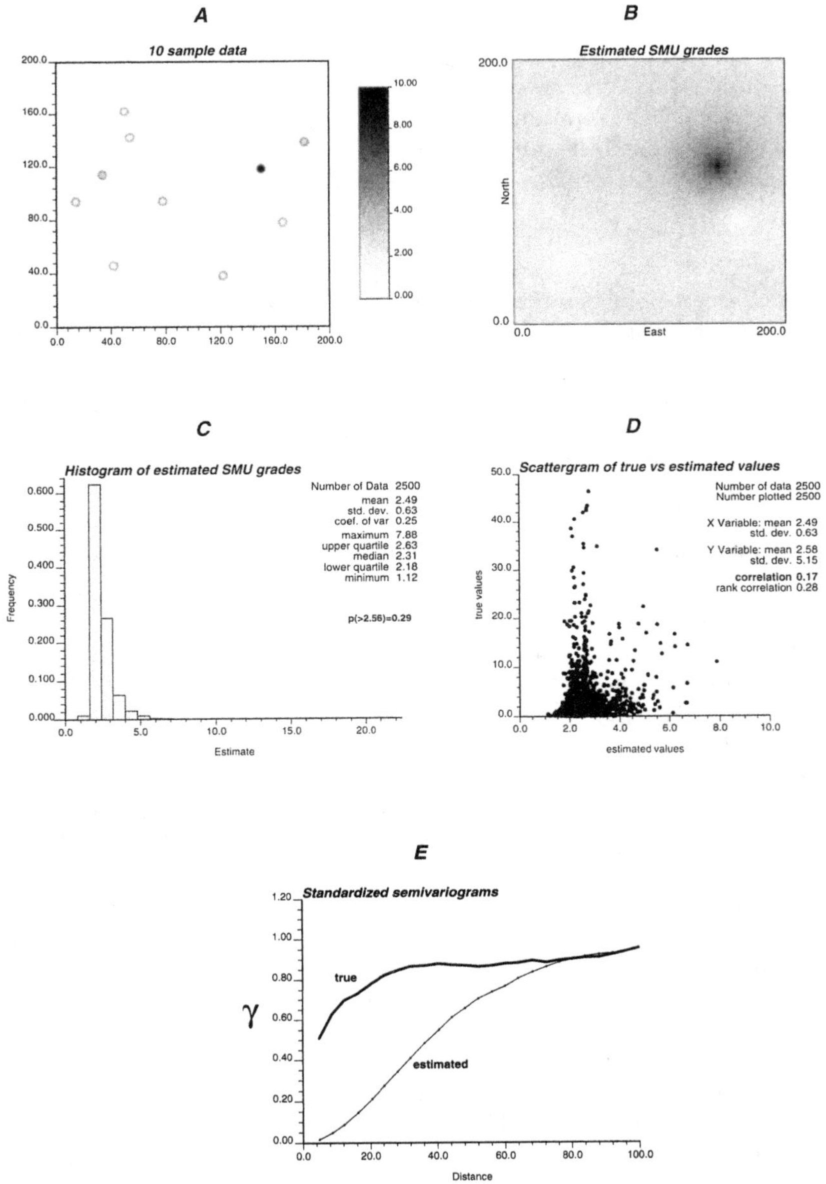

Figure I.5: Sample data and estimated SMU grades. **A**: 10 sample point support data, **B:** pixel map of estimated SMU grades, **C:** histogram of SMU estimates, **D:** scattergram of true versus estimated SMU grades, **E:** omnidirectional semivariograms of true and estimated SMU grades, standardized to their respective variances.

much smaller than the SMU volume $|v|$. The concept of conditional simulation consists of generating many alternative representations, called realizations, of the variable $z(\mathbf{u})$ with the same support as the sample data (see Journel and Huijbregts, 1978, chap. 7; or Goovaerts, 1997, chap. 8). Any such realization $\{z^{(l)}(\mathbf{u}), \mathbf{u} \in D\}$ aims at reproducing the histogram and variogram modeling the spatial distribution of the sample data available, in addition to identifying the data values at their locations, a property known as data exactitude:

$$z^{(l)}(\mathbf{u}_\alpha) = z(\mathbf{u}_\alpha), \quad \forall \alpha, l \tag{I.11}$$

Any such quasi-point support realization $\{z^{(l)}(\mathbf{u}), \mathbf{u} \in D\}$ can be generated over a very dense grid; for example, tens of nodes could be used to discretize each SMU $v(\mathbf{u})$. The point support realization can then be locally averaged into a corresponding simulated realization $\{z_v^{(l)}(\mathbf{u}), \mathbf{u} \in V\}$ of the SMU grades over any panel V. A large number $l = 1, \ldots, L$ of such realizations can be generated, which are all equally likely to be drawn resulting in an equal number L of equiprobable simulated realizations $\{z_v^{(l)}(\mathbf{u}), \mathbf{u} \in V\}$, $l = 1, \ldots, L$ of the SMU grades.

Figures I.6A and I.6B give two realizations of the SMU grades, and the statistics of the first realization are shown in Figures I.6C–E. The histogram and semivariogram of the simulated SMU grades (Figs. I.6C–E) are seen to be much closer to the corresponding statistics of the true SMU grades (Figs. I.4B and I.4C) than the statistics of the estimated SMU grades (Figs. I.5C–E). In particular, the proportion of simulated values $z_v^{(1)}(\mathbf{u}) > z_c = 2.56$ is 0.27, a value closer to the 25% of such true values. Conditional simulation, if properly performed, does not suffer from the conditional bias of estimation performed with the *same* data. The better reproduction of global and structural statistics (histogram and variogram) is obtained at the expense of poorer local accuracy: Figure I.6D reveals a greater scatter of simulated versus true values and a poorer correlation (0.04) than that (0.17) of Figure I.5D. Structural accuracy is necessarily obtained at a loss of local accuracy (e.g., see Journel et al., 2000).

If one accepts that locating recoverable SMUs is not possible from too few sample data, then the structural accuracy provided by conditional simulation suffices in that it ensures unbiasedness of the estimates of global recovery. The slightly better local accuracy provided by kriging is usually not enough to accurately locate those SMUs that will be actually recovered based on the future, and likely much larger, data set available at the time of selection.

Assessment of uncertainty

Any estimation algorithm (e.g., kriging) will result in a unique map of estimated values from a given data set (see Fig. I.5B). However, the characteristic of simulation algorithms is that they result in many (equiprobable) simulated realizations, two of which are shown in Figures I.6A and I.6B. Because all L realizations honor the same constraints (data exactitude, histogram, and

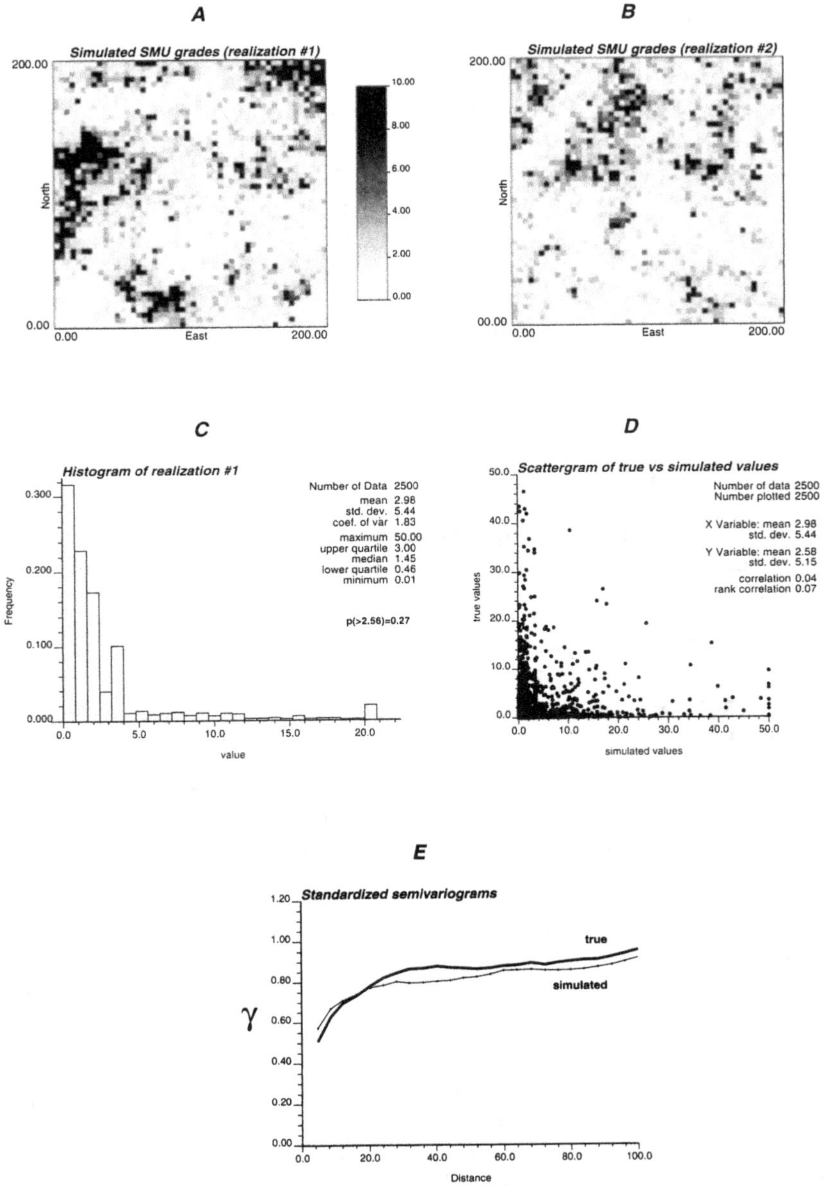

Figure I.6: Two simulated realizations (**A**, **B**) of SMU grades, and statistics of the first realization (**C**–**E**).

variogram model), where they differ and how much they differ provides a measure of uncertainty about the phenomenon being represented, which here is the spatial distribution of SMU grades.

Consider any property, say $meas(l)$, calculated from any one realization l. The histogram of the L values $\{meas(l), l = 1, \ldots, L\}$ provides a measure of uncertainty of that property. Note that for this histogram to be an unbiased measure of uncertainty, the L realizations $\{z_v^{(l)}(\mathbf{u}), \mathbf{u} \in V\}, l = 1, \ldots, L$, must be equally likely to be drawn.

Beware that any measure of uncertainty is associated with the stochastic model underlying the simulation algorithm that was used to generate the set of synthetic realizations. If that stochastic model is too favorable, the various realizations would tend to be all globally similar, thus understating uncertainty and, in some cases, failing to to bracket the "true" values. There is no universal or "objective" measure of uncertainty (see later discussion in Chapter II).

I.5 Scope of the Book

One important source of error in the evaluation of recoverable reserves and associated revenues is the ignorance or imperfect correction of both the support effect and the information effect. More precisely:

- The statistics and spatial distribution of SMU grades are different from those of core or composite data available at the time of reserves evaluation. SMU volumes are orders of magnitude larger than the volume support of drillhole data.

- SMUs are selected on the basis of their *future* estimated grades, themselves built from future data of different support and sample size than those of the data available at the time of reserves evaluation.

This book aims at presenting how the impact of these two effects (support and information) can be evaluated through simulation. From the data available at the time of reserves evaluation, both the SMU grade $z_v(\mathbf{u})$ and its future selection estimate $z_v^{**}(\mathbf{u})$ can be simulated by the pairs $\{z_v^{(l)}(\mathbf{u}), z_v^{**(l)}(\mathbf{u})\}$, thus allowing a simulation of the future recovery.

Geologists, mining engineers, and mine planners in general may discover the stochastic simulation tool: one can simulate mine planning and ore recovery in advance of opening the mine, as aeronautical engineers simulate the flight performance of a plane before it is built. Both the software (simulation algorithms) and the hardware (fast CPUs and large RAM) are now available; experience from extensive practice (in the oil industry) is also available. There remains to develop the mining practice of stochastic simulations; our hope is that this book will contribute to that development. This book is written mainly for undergraduate geostatisticians and mining engineers. It does not delve into any optimization or geostatistical theory; references are made to a few widely accessible books and the public-domain geostatistical library (GSLIB) software. Any new technology is best used when its theory is mastered: the professional desirous to try out the simulation approach would do well learning the basics of geostatistics through a short course or a semester class, or reading the introductory book by Isaaks and Srivastava (1989) followed by the more complete reference of Goovaerts (1997). All geostatistical programs used in this book were taken directly from the GSLIB software (Deutsch and Journel, 1998). Other utility programs, such as those used for global recovery and profit function calculations, can be readily written.

Geostatisticians, both experts and beginners, may profit from the detailed exposition of a particular application of geostatistical reasoning and algorithms to a real (and difficult) problem. Geostatistics rose from the need of applications—it has never been theory in the search of applications. The multiple variants of kriging and conditional simulation are best understood from an application point of view, and more variants will be developed in the future to accommodate the specifics of each data set and problem at hand. Except for the simulation of future blastholes (Section IV.3.1), no new algorithm is presented in this book. What we believe to be reasonably novel is the systematic treatment and simulation of the various aspects conditioning recovery of ore reserves.

Ideally, a real mining data set, exhaustive and clean, should have been used to provide a reference to which all estimated and simulated results could be compared. Because such a mining data set does not exist, we used instead "an exhaustive and clean" topographic data set, which was appropriately transformed to mimic the statistics of mining data. Such a real "natural" data set, although not originating from mining, would not preferentially privilege algorithms based on a prior assumption of a multivariate Gaussian distribution for the available data or, equivalently, the assumption of maximum spatial entropy ("homogeneous heterogeneity") [see Journel and Deutsch, 1993]. Students in geostatistics may consider repeating and extending the exercises of this book to other natural data sets more akin to their application fields.

All data sets used in this book are available for downloading via the Internet from the following Web site:

`http://ekofisk.stanford.edu/SCRFweb/mineral_reserves/index.html`

Chapter II

The Simulation Approach

This book is not about how to generate or simulate numerical models of a mineral deposit, but about how to use simulated numerical models of the grade spatial distribution for evaluating recoverable reserves.

In this chapter, the concept of simulation from a random function model is briefly reviewed. Simulation results in alternative, equiprobable, numerical representations of the SMU grade spatial distribution. On each such numerical representation or "realization", the process of selection and ore recovery can be itself simulated.

Section II.1 recalls the concept of a random function model and stochastic simulation from such a model. Averaging of point support simulated values yields simulated realizations on any support volume or SMU size (Section II.2). The simulation of future data and the subsequent generation of selection estimates is addressed in Section II.3. From the simulated realizations of SMU grades and their selection estimates, both simulated and estimated recovery can be retrieved (Section II.4).

Why simulation?

Actual selection is performed on estimated values based on data available at the time of selection, yet true grades are recovered. If one could know the scattergram of true SMU grades versus their selection estimates, as in Figure I.3, the problem of estimation of recoverable reserves would be solved exactly. Short of such an exhaustive knowledge, we propose to simulate that scattergram; that is, replace the pairs of unknown values $\{z_v(\mathbf{u}), z_v^{**}(\mathbf{u}), \mathbf{u} \in V\}$ within any panel V by simulated pairs $\{z_v^{(l)}(\mathbf{u}), z_v^{**(l)}(\mathbf{u}), \mathbf{u} \in V\}$. There can be L such simulated sets of values, $l = 1, \ldots, L$. The difference between the L sets of results provides an assessment of uncertainty within the simulation model used. This uncertainty can be assessed at various scales, from that of an individual panel to that of a bench or mining face or that of the whole deposit.

Important remarks

- Free selection is assumed within panel V. Any SMU $v(\mathbf{u})$ is selected on the basis of its simulated future estimated value $z_v^{**(l)}(\mathbf{u})$, independent of the status of neighboring SMUs. The panel V could be of any size, up to the dimensions of the entire final pit. However, the larger the panel V, the more unrealistic is the assumption of free selection.

- Knowledge of the set of true SMU grades $\{z_v(\mathbf{u}), \mathbf{u} \in V\}$ is not enough to determine the recoverable reserves, because SMUs are selected on the basis of estimates, *not* of true values. The information effect, that is, the impact of future misclassification, must be accounted for.

II.1 Point Grade Simulation

Because the proposed approach calls for the *joint* simulation of the true SMU value $z_v(\mathbf{u})$ and its future estimated value $z_v^{**}(\mathbf{u})$, both conditioned to the presently available data $\{z(\mathbf{u}_\alpha), \alpha \in (n)\}$, consistency is an absolute must. Such a consistency is best achieved by simulating the common variable underlying both $z_v(\mathbf{u})$ and $z_v^{**}(\mathbf{u})$, which is the variable $z(\mathbf{u})$ with the same support volume as that of the present data $z(\mathbf{u}_\alpha)$. This support is assumed to be quasi-point with regard to the SMU support v. From any simulated realization $\{z^{(l)}(\mathbf{u}), \mathbf{u} \in V\}$ of the point support variable $z(\mathbf{u})$, consistent realizations of the true SMU grade $z_v(\mathbf{u})$ and its future estimate $z_v^{**}(\mathbf{u})$ can be derived by averaging; these are denoted respectively:

$$\{z_v^{(l)}(\mathbf{u}), \mathbf{u} \in V\} \quad \text{and} \quad \{z_v^{**(l)}(\mathbf{u}), \mathbf{u} \in V\}, \quad \text{for the } l\text{th realization.}$$

Model building

Any geostatistical study (simulation is no exception) starts with a thorough analysis of the information available at the (present) time of the prediction exercise. This includes the set (n) of grade values, denoted as $\{z(\mathbf{u}_\alpha), \alpha = 1, \ldots, n\}$, defined on the quasi-point support of core samples or composite data. It also includes structural data, such as geological interpretation about rock type arrangement, directions of trends and continuity, and so forth. That information, both hard (numerical) and soft or putative (e.g., structural interpretation), should be analyzed carefully, resulting in a geostatistical model for the spatial distribution of the point support grades within panel V: $\{z(\mathbf{u}), \mathbf{u} \in V\}$.

Spatial distribution models are typically characterized by a set of histograms and variograms (or covariances). Within the panel V there may be various rock or mineralization types, whose grades are sufficiently different to warrant their separation in distinct populations (see Rendu and Ready, 1982; or David, 1988). The spatial arrangement of these different rock types may be characterized by the corresponding proportions and variograms of rock type indicator variables, see Davis (1984), Journel (1989, Lesson 4), Dimitrakopoulos and Dagbert (1993). Alternatively, the spatial arrangement of different rock types may be characterized by random geometry parameters (Haldorsen and Damsleth, 1990; Damsleth et al., 1992) or by multiple-point statistics inferred from training images (Strebelle, 2002). Estimation or simulation of the geometry of the mineralization zones is as critical as the simulation of the corresponding metal grades. Within each zone or rock type, the distribution of the grade $z(\mathbf{u})$ is characterized by a distinct histogram and variogram. All these statistics must be gathered from the information presently available and allow building a random function (RF) model for the spatial distribution of $z(\mathbf{u})$ within the panel V.

A random function, denoted with a capital letter Z as $\{Z(\mathbf{u}), \mathbf{u} \in V\}$, can be seen as a set of spatially correlated random variables $Z(\mathbf{u})$, one per location $\mathbf{u}$. A random variable (RV) $Z(\mathbf{u})$ can be seen as the set of all possible outcomes of the unknown value $z(\mathbf{u})$ ranked by the cumulative distribution function (cdf):

$$F(\mathbf{u}; z) = \text{Prob}\{Z(\mathbf{u}) \leq z\}$$

read as: "probability for an outcome to be valued less than or equal to threshold z".

Stochastic simulation

Various techniques exist for drawing realizations from a RF model $\{Z(\mathbf{u}), \mathbf{u} \in V\}$; that is, joint realizations of its constituent RVs $Z(\mathbf{u})$. One realization of the RF $\{Z(\mathbf{u}), \mathbf{u} \in V\}$ constitutes one stochastic representation or image of the possible spatial distribution of the point grades $z(\mathbf{u})$ within panel V; that realization is denoted as $\{z^{(l)}(\mathbf{u}), \mathbf{u} \in V\}$, with the superscript (l) indicating the realization number.

There can potentially be an infinite number of such realizations $\{z^{(l)}(\mathbf{u}), \mathbf{u} \in V\}$; in practice a finite number L is generated, with typically $L = 50$ or 100. All realizations are equally likely to be drawn. They are all conditional to the present numerical data, in that:

$$z^{(l)}(\mathbf{u}_\alpha) = z(\mathbf{u}_\alpha), \quad \forall\alpha \in (n), \ \forall l = 1, \ldots, L, \quad \text{recall relation (I.11)}$$

They all honor (reproduce) the prior structural information typically limited to the prior models of histograms and variograms; this is done within statistical fluctuations from one realization l to another l'.

There are numerous publications dealing with structural analysis (e.g., Isaaks and Srivastava, 1989, chaps. 2–7; and Deutsch, 2002), random function modeling and simulation (e.g., Goovaerts, 1997, chaps. 3–4); refer also to the related public domain geostatistical software GSLIB in Deutsch and Journel (1998).

At this point we have available L simulated realizations $\{z^{(l)}(\mathbf{u}), \mathbf{u} \in V\}$, $l = 1, \ldots, L$, of the point support grade spatial distribution within the panel V. These simulated values are not used directly for estimation of recoverable reserves because they do not appear in the scattergram of Figure I.3. Instead, they are used to provide a *coherent* joint simulation of both SMU grades $z_v(\mathbf{u})$ and their final estimated values $z_v^{**}(\mathbf{u})$.

II.2 Simulation of SMU Grades

If the simulation of the point support grades $\{z^{(l)}(\mathbf{u}), \mathbf{u} \in V\}$ is done on a grid sufficiently dense to discretize each SMU $v(\mathbf{u})$ by n_v points, the simulated SMU averaged grade $z_v^{(l)}(\mathbf{u})$ can be approximated by the arithmetic (linear) average of the n_v simulated point values within $v(\mathbf{u})$:

$$z_v^{(l)}(\mathbf{u}) = \frac{1}{|v|} \int_{v(\mathbf{u})} z^{(l)}(\mathbf{u}')d\mathbf{u}' \simeq \frac{1}{n_v} \sum_{j=1}^{n_v} z^{(l)}(\mathbf{u}'_j) \tag{II.1}$$

In practice, a discretization leading to $n_v = 10 \times 10$ for a two-dimensional area or $n_v = 10 \times 10 \times 4$ for a three-dimensional volume is amply sufficient for a mineralization characterized by a variogram with not too large nugget effect (less than 30%). The final goal is not a high-resolution representation of the point support grade spatial distribution, which would be highly uncertain, but a representation of the much smoother SMU support grade spatial distribution. For a panel V that comprises 1000 SMUs, the previous three-dimensional discretization would lead to a point support simulation grid with 4×10^5 nodes; simulation on such a grid is well within the capacity of desktop computers available at the time of editing this book (2003). Simulation on a grid of several million nodes, corresponding to a large panel V, would still be possible with enough central memory to store one realization at a time for fast postprocessing. It would not be reasonable to attempt simulation of a full deposit at the point support scale; issues of stationarity (zoning) would most often call for splitting that deposit into several domains/levels/benches.

The resulting simulated SMU support grade realizations, $\{z_v^{(l)}(\mathbf{u}), \mathbf{u} \in V\}$, $l = 1\ldots, L$, reflect the spatial distribution of the true SMU grades $z_v(\mathbf{u})$, in that they reproduce the variogram of the SMU grades defined as:

$$2\gamma_v(\mathbf{h}) = E\{[Z_v(\mathbf{u}) - Z_v(\mathbf{u}+\mathbf{h})]^2\}$$

This SMU support variogram $2\gamma_v(\mathbf{h})$ is fully determined by the point support variogram model $2\gamma(\mathbf{h})$ used for the generation of the point support grade realizations $\{z^{(l)}(\mathbf{u}), \mathbf{u} \in V\}$. For a proof, the reader should refer to publications on the topic of regularization (e.g., Journel and Huijbregts, 1978, p. 77; Deutsch, 2002).

There is no data exactitude because of the difference of support between the SMUs and the data $z(\mathbf{u}_\alpha)$. Conditional simulation, however, will ensure that all SMU grades $z_v^{(l)}(\mathbf{u})$ simulated in the vicinity of a high-grade datum $z(\mathbf{u}_\alpha)$ will be "similarly" high. How similar depends on the variogram model $2\gamma(\mathbf{h})$ used, and particularly on its nugget effect.

Direct simulation of SMU grades

An alternative to the averaging (II.1) of simulated point grades into simulated SMU grades is to simulate directly the SMU values $z_v^{(l)}(\mathbf{u})$ conditional to the quasi-point support data $\{z(\mathbf{u}_\alpha), \alpha \in (n)\}$. This would call for a prior model of the SMU grade histogram [see block kriging in Journel and Huijbregts (1978, p. 304) and block simulation in Deutsch and Journel (1998, p. 296)]. Such direct simulation would be more CPU- and much more RAM-efficient, but it requires that all SMUs be of exactly the same size $|v|$. Were the study to require another SMU size $|v|' \neq |v|$ to be considered, the simulation exercise would have to be repeated. This is not so if one starts by simulating point grades $z^{(l)}(\mathbf{u})$, because these can be linearly averaged into SMU grades of any size (with that size possibly varying in space).

The point support approach is recommended; that is, simulating values $z^{(l)}(\mathbf{u})$ of the same support as the data $z(\mathbf{u}_\alpha)$ available, followed by the averaging (II.1). This approach ensures consistency between simulated point values and block values defined on any larger support v, at the cost of greater memory requirements.

SMU grade histogram

The histogram of the simulated SMU grades $z_v^{(l)}(\mathbf{u})$ within any panel V is a direct result of the averaging (II.1). Actually, each realization (l) yields a different histogram of simulated values $\{z_v^{(l)}(\mathbf{u}), \mathbf{u} \in V\}$. If the dimensions of panel V are not large with regard to the range(s) of the variogram model $2\gamma(\mathbf{h})$ used, the fluctuations of these different SMU support histograms could be significant. The average of the L simulated SMU support histograms provides an estimate of the unknown true SMU grade histogram, and their fluctuations provide an assessment of uncertainty regarding that estimate.

If one proceeds, as recommended above, with point support simulation averaged later into simulation of block (SMU) grades, one may naively think that no prior hypothesis about the block grade probability distribution (histogram) has been made. This is not so, because the simulation of point values within any panel V requires, implicitly or explicitly, the multivariate probability distribution of point values within V, which includes precisely the block probability distributions (see related discussions in Deutsch and Journel, 1998, p. 92–94).

Thus, proceeding with point support simulation does not exempt one from a block distribution hypothesis—that hypothesis is simply embedded in the random function model $\{Z(\mathbf{u}), \mathbf{u} \in V\}$ used for the simulation of the point support grades. Again, the main advantage of proceeding with point support simulation is consistency of all results, a particularly critical advantage if different SMU sizes $|v|' \neq |v|$ are to be considered and the corresponding recoveries compared.

II.3 Simulation of Selection Criteria

The objective now is to deduce from the point support simulated realization, $\{z^{(l)}(\mathbf{u}),\ \mathbf{u} \in V\}$, the realization of the final (future) selection estimates $\{z_v^{**(l)}(\mathbf{u}),\ \mathbf{u} \in V\}$. Indeed, actual selection will be performed on such future estimated values based on the data available at the time of mining—recall the discussion related to Figure I.3.

The first step is to simulate the future selection data, denoted as $\{z^{(l)}(\mathbf{u'}_\alpha),\ \alpha = 1, \ldots, nn\}$, $l = 1, \ldots, L$, with typically $nn >> n$. Each such realization, say $z^{(l)}(\mathbf{u'}_\alpha)$, must be consistent with the point support simulated values $z^{(l)}(\mathbf{u})$ in its neighborhood for the same realization identification number l.

The second step is to generate future selection estimates $\{z_v^{**(l)}(\mathbf{u}), \mathbf{u} \in V\}$ from each realization of simulated future data $\{z^{(l)}(\mathbf{u'}_\alpha),\ \alpha = 1, \ldots, nn\}$. This second step is a mere (deterministic) estimation step mimicking the future estimation of SMU grades from selection data (e.g., blasthole grades). For example, the estimated SMU values $z_v^{**(l)}(\mathbf{u})$ could be obtained by block kriging from the selection data $z^{(l)}(\mathbf{u'}_\alpha)$:

$$z_v^{**(l)}(\mathbf{u}) = \sum_{\alpha=1}^{nn(\mathbf{u})} \lambda_\alpha(\mathbf{u}) z^{(l)}(\mathbf{u'}_\alpha) \tag{II.2}$$

where $nn(\mathbf{u})$ is the number of future data retained in the neighborhood of, including within, the SMU $v(\mathbf{u})$, and the estimation weights $\lambda_\alpha(\mathbf{u})$ are given by kriging or any other interpolation procedure as long as it mimics that used at the time of actual selection. For block kriging, refer to Goovaerts (1997, p. 152).

Simulation of future data

Simulation of future data from realizations $\{z^{(l)}(\mathbf{u}), \mathbf{u} \in V\}$ of quasi-point support values is trivial if the future selection data are assumed of the same type (support and quality) as the present data $z(\mathbf{u}_\alpha)$, on which the simulated values $z^{(l)}(\mathbf{u})$ are conditioned. For example, if the present data $z(\mathbf{u}_\alpha)$ are core data, the future selection data $z(\mathbf{u'}_\alpha)$ would have to be also core data. It suffices then to retain from the dense grid of point support simulated grades $\{z^{(l)}(\mathbf{u}), \mathbf{u} \in V\}$ the subset of values $\{z^{(l)}(\mathbf{u'}_\alpha),\ \alpha = 1, \ldots, nn\}$ corresponding to the expected future grid of selection data.

The problem is much more difficult if the future selection data $z(\mathbf{u'}_\alpha)$ are of different type than the presently available data $z(\mathbf{u}_\alpha)$. In such a case, we should start using different notations for the future selection data, say:

$$\{b(\mathbf{u'}_\alpha), \alpha = 1, \ldots, nn\} \tag{II.3}$$

keeping the notation $\{z(\mathbf{u}_\alpha), \alpha = 1, \ldots, n\}$ for the present data on which the point support realizations $\{z^{(l)}(\mathbf{u}), \mathbf{u} \in V\}$ are based.

Examples are, in an open pit setting, blasthole data $b(\mathbf{u'}_\alpha)$ versus present diamond drillhole (core) data $z(\mathbf{u}_\alpha)$, and in an underground setting, reverse circulation or longhole data or channel sample data $b(\mathbf{u}'_\alpha)$ versus present core samples $z(\mathbf{u}_\alpha)$.

To be able to simulate the future data $b(\mathbf{u}'_\alpha)$ consistently with the simulation $z^{(l)}(\mathbf{u})$, one must know the statistics relating the two types of data; that is, the statistics relating the two random functions $\{B(\mathbf{u}), \mathbf{u} \in V\}$ and $\{Z(\mathbf{u}), \mathbf{u} \in V\}$. Inference of such joint statistics is a most difficult task, as, by definition, at the time of prediction no future selection data $b(\mathbf{u}'_\alpha)$ are yet available. From prior experience on similar selective mining operations on different parts of the same operating mine or on different mines, one must infer the statistics of the future data $b(\mathbf{u}'_\alpha)$ [e.g., Pitard, 1993].

One way to ensure consistency is to simulate the future data $b^{(l)}(\mathbf{u}'_\alpha)$ by adding to the simulated value $z^{(l)}(\mathbf{u}'_\alpha)$ at the same location a simulated error or deviation $r^{(l)}(\mathbf{u}'_\alpha)$:

$$b^{(l)}(\mathbf{u}'_\alpha) = z^{(l)}(\mathbf{u}'_\alpha) + r^{(l)}(\mathbf{u}'_\alpha), \quad \forall \alpha = 1, \ldots, nn, \; \forall l$$

or in terms of RF models:

$$B(\mathbf{u}) = Z(\mathbf{u}) + R(\mathbf{u}), \quad \forall \mathbf{u} \in V \tag{II.4}$$

Expression (II.4), which models the poorer-quality future data as the better-quality data plus an error, may appear deceptively trivial. Actually, any two variables $B(\mathbf{u})$ and $Z(\mathbf{u})$ can be related through expression (II.4) with $R(\mathbf{u}) = B(\mathbf{u}) - Z(\mathbf{u})$. The challenge is to figure out the correlation between the error or deviation $R(\mathbf{u})$ and the "true" signal $Z(\mathbf{u})$. Too often, the error $R(\mathbf{u})$ is assumed to be independent of the signal $Z(\mathbf{u})$, an "homoscedastic" situation extremely congenial and highly unrealistic, at least when applied to sampling errors (see Goovaerts, 1997, p. 166, 261; and Gy, 1998). Instead, not only the error variance but also the error distribution type (generally non-Gaussian) depends on the underlying signal value. When sampling from blasthole or reverse circulation cuttings, the relative error $r(\mathbf{u})/z(\mathbf{u})$ typically increases at both ends of the signal $z(\mathbf{u})$ range: at the low end because of detection limits, for example, and at the high end because of loss of fines (Springett, 1988).

Ideally, actual data $\{b(\mathbf{u}'_\alpha), z(\mathbf{u}'_\alpha)\}$ from a similar mining operation would allow inference of the statistics of the error RF $\{R(\mathbf{u}), \mathbf{u} \in V\}$ and its simulation conditional to the previously generated signal values $\{z^{(l)}(\mathbf{u}), \mathbf{u} \in V\}$. Otherwise, different scenarios for the distribution (statistics) of the deviation $R(\mathbf{u})$ must be considered. Each different scenario would lead to a different realization of the future selection data for any given realization of the signal $\{z^{(l)}(\mathbf{u}), \mathbf{u} \in V\}$. An optimistic scenario would consist of considering the future data $b(\mathbf{u}'_\alpha)$ of the same type and quality as the present data $z(\mathbf{u}'_\alpha)$; this is the case considered later in Section V.3. An example of modeling and simulation of blasthole data of different type than present (core) data is given in Section IV.3.1.

Note that there are cases where the future grade control data are of better quality than the presently available development data. One would then reverse the notations B and R in expression II.4, defining the error from the better-quality data as:

$$Z(\mathbf{u}) = B(\mathbf{u}) + R(\mathbf{u})$$

II.4 Simulation of Recovery Functions

At this point we have all the (simulated) data necessary to figure out the recoverable reserves assuming: (i) free selection over the panel V, (ii) selection performed on future estimated values of the type $z_v^{**}(\mathbf{u})$, and (iii) actual recovery of the corresponding true SMU grades $z_v(\mathbf{u})$.

Taking each simulated realization $\{z_v^{(l)}(\mathbf{u}), z_v^{**(l)}(\mathbf{u}), \mathbf{u} \in V\}$ as a possible outcome of the actual but unknown pair $\{z_v(\mathbf{u}), z_v^{**}(\mathbf{u}), \mathbf{u} \in V\}$, one can apply the recovery formulae (I.6) to (I.10). For example, the lth simulated realization of the "actual" selection indicator at cutoff $z = z_c$ is:

$$i_v^{**(l)}(\mathbf{u}; z_c) = \begin{cases} 1 & \text{if } z_v^{**(l)}(\mathbf{u}) > z_c \\ 0 & \text{if not} \end{cases} \tag{II.5}$$

The corresponding lth simulated realizations of the "actual" recovery functions are:

$$\text{Ore tonnage:} \quad t_v^{(l)}(z_c) = \sum_{j=1}^{N} i_v^{**(l)}(\mathbf{u}_j; z_c) \tag{II.6}$$

$$\text{Quantity of metal:} \quad q_v^{(l)}(z_c) = \sum_{j=1}^{N} i_v^{**(l)}(\mathbf{u}_j; z_c) \cdot z_v^{(l)}(\mathbf{u}_j)$$

$$\text{Mean ore grade:} \quad m_v^{(l)}(z_c) = q_v^{(l)}(z_c)/t_v^{(l)}(z_c)$$

$$\text{Profit:} \quad p_v^{(l)}(z_c) = \text{a known function of } t_v^{(l)}(z_c), q_v^{(l)}(z_c)$$

In these definitions, the qualifier "actual" recalls that the target recovery to be estimated is recovery based on actual selection performed on future SMU estimated values, as opposed to "perfect" recovery based on perfect selection performed on true SMU grades.

Uncertainty assessment and estimation

Typical of modern geostatistics, assessment of uncertainty about the unknown recovery values is done before, and independent of, the particular estimated value retained.

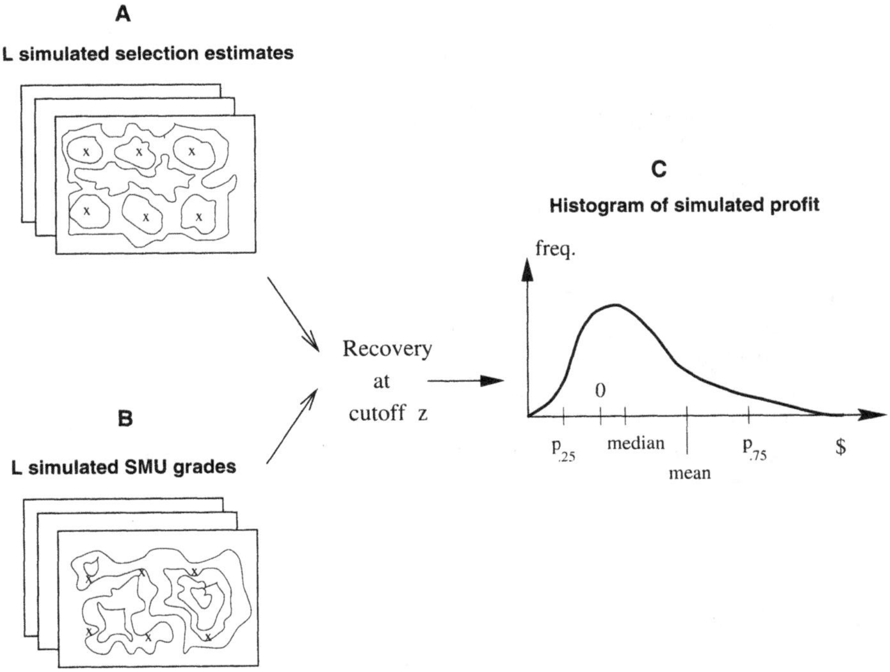

Figure II.1: Assessment of uncertainty about actual profit. SMU selection is performed on simulated future estimates (**A**), with the corresponding simulated true SMU grades (**B**) being recovered. These L realizations of SMU-related values yield L simulated profit values, whose histogram is shown in (**C**). From that histogram, probability intervals can be retrieved, such as the 50% probability interval $[p_{.25}, p_{.75}]$, and "best" estimates for the profit. The least squared error best estimate is the mean simulated profit.

Consider, for example, the histogram of the L simulated actual profit values $\{p_v^{(l)}(z_c), l = 1, \ldots, L\}$ at cutoff z_c, shown in Figure II.1C:

- The interval bounded by the 25th and the 75th percentiles of that histogram gives an estimate of the 50% probability interval for the actual (unknown) profit:

$$\text{Prob}\{p_v^{(l)}(z_c) \in (p_{.25}(z_c), p_{.75}(z_c)]\} = 0.5 \tag{II.7}$$

  where, for example, $p_{.25}(z_c)$ is the lower quartile value of the simulated profit histogram.

  Note that these probability intervals do not call for any arbitrary assumption of Gaussian distribution of errors. Also, they are not attached to any particular estimated value $[p_v(z_c)]^*$ for the profit recovered at z_c.

- For any specific optimality criterion, the "best" estimate $[p_v(z_c)]^*$ can be retrieved. For example, if that criterion is the traditional least squared error criterion, then the best estimate would be the mean of the L simulated profit values (see Fig. II.1):

$$[p_v(z_c)]^*_{LS} = \frac{1}{L}\sum_{l=1}^{L} p_v^{(l)}(z_c) \tag{II.8}$$

  If one wishes to retain the estimate that minimizes the mean absolute error, however, the best estimate is the median $p_{.50}(z_c)$ of the L simulated profit values, a value possibly significantly different from the previous mean $[p_v(z_c)]^*_{LS}$. Yet, no matter how different these two "best" estimates are, the previous probability interval (II.7) for the profit remains the same. For an extended discussion on this important topic, the reader is referred to Srivastava (1987) or Goovaerts (1997, p. 340).

Misclassification
Availability of the L simulated realizations $\{z_v^{(l)}(\mathbf{u}), z_v^{**(l)}(\mathbf{u}), \mathbf{u} \in V\}$ allows an assessment of the level of misclassification and related economic impact, associated with the degree of selectivity assumed. By degree of selectivity, it is understood the support size $|v|$ of the SMUs and the information (nn) available at the time of selection.

Indeed, each simulated realization $\{z_v^{(l)}(\mathbf{u}), z_v^{**(l)}(\mathbf{u}), \mathbf{u} \in V\}$ allows generation of a simulated classification scattergram of the type of Figure I.3, from which misclassification statistics can be retrieved (statistics of sectors III and IV in Fig. I.3). Sensitivity analyses can be performed by varying the type and density of data available for future selection: only the values $z_v^{**(l)}(\mathbf{u})$ would change; or varying the SMU size $|v|$: both values $z_v^{(l)}(\mathbf{u})$ and $z_v^{**(l)}(\mathbf{u})$ would change. One could then balance the cost of a more accurate selective mining (smaller SMU v, more or better selection data) with the greater revenue achieved from less misclassification leading to increased metal recovery. Again, such an optimization study is predictive; it is based on the presently available grade data (n), before actual mining and selection.

II.5 Reflections

In the presence of uncertainty, simulation represents an approach fundamentally different from the more traditional estimation.

In estimation, one seeks a unique and "best" estimate for each unknown value; for example, an estimate $z_v^*(\mathbf{u})$ of a particular SMU grade $z_v(\mathbf{u})$. By "best" it is generally understood that value $z_v^*(\mathbf{u})$, which postmortem, would minimize the average squared error $\frac{1}{N}\sum_{j=1}^{N}[z_v^*(\mathbf{u}_j) - z_v(\mathbf{u}_j)]^2$ taken over N similarly estimated values. The potential error interval associated with the specific estimate $z_v^*(\mathbf{u})$ at location $\mathbf{u}$ is rarely evaluated. Instead, and only if geostatistics is called on, the calculated error variance (e.g., the kriging variance) is actually an evaluation of the previous spatially averaged squared error (Journel, 1985).

The problem with estimation, particularly in the presence of sparse data, is that the set (map) of estimated values $\{z_v^*(\mathbf{u}_j), j = 1, \ldots, N\}$ does not reflect the actual heterogeneity or, equivalently, the spatial structure of the true values $\{z_v(\mathbf{u}_j), j = 1, \ldots, N\}$; this is the well-known smoothing effect of estimation. In other words, even if estimated values $z_v^*(\mathbf{u}_j)$ achieve the best *local* accuracy possible given the available data, they do not deliver *global* accuracy. The problem is that prediction of recoverable reserves calls for an accurate representation of global statistics, such as a map or a histogram of SMU grades. Prediction of recoverable reserves *does not* call for accurate localization of each SMU that will be mined as ore in the future. One must distinguish the task of early prediction of recoverable reserves for mine feasibility (the subject of this book) from that of ore grade control; that is, accurate localization of ore stopes at the time of mining. The former task requires in priority global and structural accuracy as provided by simulation, the latter task calls for local accuracy as provided by estimation. Short of exhaustive sampling, one cannot achieve simultaneously both types of accuracy (Journel et al., 2000).

In simulation, one seeks an accurate global representation of the spatial distribution of the SMU grades $z_v(\mathbf{u})$. The spatial continuity and histogram of the simulated SMU grades $z_v^{(l)}(\mathbf{u})$ should mimic that of the yet unknown true values $z_v(\mathbf{u})$. Indeed, a correct assessment of the spatial continuity of ore zones is important for mine planning (accessibility, selectivity), and a correct assessment of the histogram of SMU grades (particularly its upper tail) is essential for evaluating recovery of the best grades. In simulation, one is prepared to trade some local accuracy for better structural accuracy: a kriging estimate $z_v^*(\mathbf{u})$, or equivalently, the average of the L simulated realizations $\frac{1}{L}\sum_{l=1}^{L} z_v^{(l)}(\mathbf{u})$, are more locally accurate estimates of the unknown true value $z_v(\mathbf{u})$, but a map of kriging estimates does not provide an accurate representation of the *spatial* distribution of the actual true SMU grades. A map of simulated values, say $\{z_v^{(l)}(\mathbf{u}_j), j = 1, \ldots, N\}$ for a given realization l, is more structurally accurate in that it reproduces the variogram model characterizing the spatial correlation between any two true values $z_v(\mathbf{u})$ and $z_v(\mathbf{u}+\mathbf{h})$. Advanced simulation algorithms (e.g., Strebelle, 2002), aim at reproducing whole structural patterns involving many more than only two locations $\mathbf{u}$ and $\mathbf{u}+\mathbf{h}$.

Simulation provides an additional bonus, a measure of both local and global uncertainty. For each SMU $v(\mathbf{u})$, the set of L simulated values $\{z_v^{(l)}(\mathbf{u}), l = 1, \ldots, L\}$ provides a distribution (histogram) modeling local uncertainty about any specific unknown SMU value $z_v(\mathbf{u})$. This is as opposed to the mere kriging or estimation variance provided by estimation algorithms. But much more importantly, the set of L maps of simulated values $\{z_v^{(l)}(\mathbf{u}_j), j = 1, \ldots, N\}$, $l = 1, \ldots, L$, provides an assessment of global uncertainty, say about the ore tonnage associated with the set of SMUs with grades $z_v(\mathbf{u}) > z_c$ above a given cutoff grade z_c. A set of local error variances, say kriging variances, cannot be used to assess the potential error involved in the evaluation of global reserves recovery.

Simulation allows addressing both the support and information effect on ore recovery. Simulation should be performed on the smallest support, typically the volume support of the composite grades available at the time of the prediction study. These quasi-point simulated values can then be averaged into simulated SMU grade values, with SMUs of possibly different size.

As for the information effect, one can simulate future data different from the present data, yet correlated with the latter. For example, evaluation of the effect of inaccurate selection of ore blocks requires the joint simulation of the true SMU grades $z_v(\mathbf{u})$ and their future selection estimates $z_v^{**}(\mathbf{u})$ by the pairs $\{z_v^{(l)}(\mathbf{u}), z_v^{**(l)}(\mathbf{u})\}$. The simulation of the future selection estimates $z_v^{**(l)}(\mathbf{u})$ requires a prior simulation of the future selection data, say blasthole data that are different but correlated with the drillhole data available at the time of the prediction study.

The availability of the simulated pairs $\{z_v^{(l)}(\mathbf{u}), z_v^{**(l)}(\mathbf{u})\}$ amounts to simulating reconciliation data and the corresponding mining dilution factors. The advantage of simulation is that it makes explicit the development of these dilution factors; of course this advantage is real only if the simulation process succeeds in representing accurately the global/spatial distribution of the actual SMU grade pairs $\{z_v(\mathbf{u}), z_v^{**}(\mathbf{u})\}$.

Chapter III

The Volume-Variance Correction Alternative

Because any mining venture investment requires an estimate of recoverable reserves and associated profits, many different approaches to such estimation have been developed. It would take years of compilation of great works and lesser manuals to write the history of mine valuation techniques starting from Agricola's de Re Metallica (Agricola, 1950). Such a survey would go far beyond the scope of this book, which is limited to the presentation of the stochastic simulation approach.

It is difficult, however, to present a relatively new approach in a vacuum; one needs a yardstick for comparison: How would an alternative approach fare under exactly the same conditions? Then, which alternative should one retain? Ideally, that alternative should be widely used and understood by all, yet it should be based on a clear methodology devoid of too many case-specific implementation twists. We chose to retain the volume-variance correction approach, denoted as VVC and developed hereafter in Sections III.2 and III.3. The reader is referred to David (1972), David (1988), Bideaux (1990), Rossi and Parker (1994), Srivastava (1994a), and Farrelly and Dimitrakopoulos (2002) for reviews of other approaches used in the mining industry.

Empirical approaches

Empirical approaches to the estimation of recoverable reserves are still widely used, and often very successfully so (Rossi and Parker, 1994). They are custom-designed to each particular mineralization and type of mining, and when based on common sense and well-understood prior experience, they are difficult to match. Such approaches, however, do not rely on any single formalism or methodology, and hence cannot be generalized away from the specific mining operation(s) on which they were fine-tuned.

The starting point is usually a map (or volume) of estimated in situ geological resources. This estimation is typically done by interpolation from geological sections and levels, independent of any consideration about SMU support size. In the best cases, the interpolation procedure would account first for the continuity of rock or mineralization types before proceeding with grade estimation. There is no numerical assessment of uncertainty besides a regulatory classification of reserves into proven, probable, or possible.

A mining plan is then built from such a map of in situ resources, delineating the limits of what will be excavated either as ore or waste. Next, within this mining plan, the amount of recoverable reserves (ore) is evaluated. The accessibility effect is explicitly accounted for; as for the support effect (SMU size) and information effect (data available for selection), they are lumped together into empirical "mining dilution and recovery factors". These factors reduce in situ resources into recoverable reserves within the planned mining areas. From these estimates of reserves and the corresponding estimates of profit, the initial mining plan may be reconsidered, starting an iteration process. Again, there is no general methodology, and the quality of the final mine plan much depends on the craft of the mine planners.

A special mention should be made of the rigorous modeling of the accessibility effect brought by optimal pit design (Lerchs and Grossmann, 1965; Matheron, 1984; Whittle, 1997). Unfortunately, this design does not account jointly for the support and information effect. In addition, such rigorous and analytical optimal design has no counterpart for underground mine planning.

III.1 Indicator Kriging

The volume-variance correction (VVC) actually consists of only one step; namely, the construction of probability distributions for the unknown SMU grades within a given panel V. The original step is indicator kriging (IK) or any other algorithm to obtain probability distributions for unknown grades defined on the volume support of the data available, usually a composite grade volume assimilated to a quasi-point (see Parker and Switzer, 1975; Journel, 1983; Isaaks and Srivastava, 1989, chap. 18). The quasi-point support probability distributions obtained from IK are then corrected for the support effect going from point to SMU volume, and this is where VVC comes in.

Indicator kriging is possibly the simplest tool for probabilistic inference, using observed proportions to infer probabilities for unsampled values. Consider an unsampled grade value $z(\mathbf{u})$ surrounded by n related data values $\{z(\mathbf{u}_\alpha), \alpha = 1, \ldots, n\}$ defined on the same support volume. We do not know the exact value $z(\mathbf{u})$, yet we are not totally ignorant about it, precisely thanks to the data values $z(\mathbf{u}_\alpha)$. One could accept that the value $z(\mathbf{u})$ is somewhat within the range defined by these n data values, or that the probability distribution for that unknown value $z(\mathbf{u})$ is the histogram of these n data: If a larger number of data values fall into a specific class $[z_{k-1}, z_k]$, then the unknown value $z(\mathbf{u})$ is also more likely to fall into that class. We have just introduced the most elementary indicator kriging, whereby the cumulative distribution function of the random variable $Z(\mathbf{u})$ modeling our uncertainty about $z(\mathbf{u})$ is taken as the equal-weighted proportion of data values:

$$F(z|(n)) = \text{Prob}\{Z(\mathbf{u}) \leq z|(n)\} = \frac{1}{n}\sum_{\alpha=1}^{n} i(\mathbf{u}_\alpha; z) \qquad \text{(III.1)}$$

with:

$$i(\mathbf{u}_\alpha; z) = \begin{cases} 1 & \text{if } z(\mathbf{u}_\alpha) \leq z \\ 0 & \text{if not} \end{cases}$$

A very small step up in our understanding of IK consists of accepting that some data $z(\mathbf{u}_\alpha)$ closer to the unsampled location $\mathbf{u}$ should receive more weight than data farther away. The previous equal weighted $(1/n)$ proportion is then replaced by an unequal weighted proportion. These weights could be made inversely proportional to the distance $|\mathbf{u}_\alpha - \mathbf{u}|$ between each datum location $\mathbf{u}_\alpha$ and the unsampled location $\mathbf{u}$, or better, be given by a kriging system accounting for the spatial correlation between the two indicator variables $I(\mathbf{u}_\alpha; z)$ and $I(\mathbf{u}; z)$; we have just defined the indicator kriging value for the conditional probability:

$$F(\mathbf{u}; z|(n)) = \text{Prob}\{Z(\mathbf{u}) \leq z|(n)\} = \sum_{\alpha=1}^{n} \lambda_\alpha(\mathbf{u}; z) \cdot i(\mathbf{u}_\alpha; z) \qquad \text{(III.2)}$$

where the weights $\lambda_\alpha(\mathbf{u}; z)$, specific to location $\mathbf{u}$ and threshold value z, are given by a kriging system (see following).

In expression (III.2), as opposed to (III.1), we made the conditional cumulative distribution function (ccdf), $F(\cdot)$, both a function of the location $\mathbf{u}$ and the n data values, through the notation $|(n)$, to insist that this probability is specific to the unknown at location $\mathbf{u}$ and to the n data values used. Also in expression (III.2), we did not use the superscript $(^*)$ for an estimate, because the IK value is not an estimate of some "true" probability value; all probability values for an unsampled grade are models for the uncertainty about that true grade; the IK value (III.2) is but one such model among many other possible models.

The conditional probability (III.2) can be seen as the conditional expectation of the indicator (binary) random variable $I(\mathbf{u}; z)$; it is a model for our uncertainty about whether the true grade $z(\mathbf{u})$ is greater (or not) than the threshold value z:

$$E\{I(\mathbf{u}; z)|(n)\} = [1 \times \text{Prob}\{Z(\mathbf{u}) \leq z|(n)\}] + [0 \times \text{Prob}\{Z(\mathbf{u}) > z|(n)\}]$$

From least squares theory, such expected value is best estimated by a combination of the indicator data that minimizes the expected square error or estimation variance. The least squares criterion is exactly that underlying indicator kriging (see Goovaerts, 1997, p. 284). The IK weights are given by a kriging system, given here under its ordinary kriging version:

$$\begin{cases} \sum_{\beta=1}^{n} \lambda_\beta(\mathbf{u}; z)\gamma_I(\mathbf{u}_\beta - \mathbf{u}_\alpha; z) + \mu(\mathbf{u}; z) = \gamma_I(\mathbf{u}_\alpha - \mathbf{u}; z), \quad \alpha = 1, \ldots, n \\ \sum_{\beta=1}^{n} \lambda_\beta(\mathbf{u}; z) = 1 \end{cases} \tag{III.3}$$

where $2\gamma_I(\mathbf{h}; z) = E\{[I(\mathbf{u}; z) - I(\mathbf{u} + \mathbf{h}; z)]^2\}$ is the indicator variogram at threshold z, to be modeled from the corresponding indicator data, and $\mu(\mathbf{u}; z)$ is the Lagrange parameter associated with the last equation of system (III.3).

Median IK

To each different cutoff z corresponds a different indicator random function $\{I(\mathbf{u}; z),\ \mathbf{u} \in V\}$ and a different indicator semivariogram $\gamma_I(\mathbf{h}; z)$. Thus, if K threshold values $\{z_k, k = 1, \ldots, K\}$ are retained for discretizing the ccdf $F(\mathbf{u}; z|(n))$, K different indicator semivariograms must be modeled and K different kriging systems of type (III.3) must be solved at each unsampled location $\mathbf{u}$.

A significant simplification is provided by median indicator kriging, where:

- All K indicator correlograms $\{\rho_I(\mathbf{h}; z_k), k = 1, \ldots, K\}$ are modeled by the same function $\rho(\mathbf{h})$, usually inferred from the indicator data at the median threshold value z_M or from the z-data themselves:

$$\rho_I(\mathbf{h}; z_k) = \rho(\mathbf{h}), \quad \forall z_k \tag{III.4}$$

- Consequently, if the same n data $\{z(\mathbf{u}_\alpha), \alpha = 1, \ldots, n\}$ are retained to derive the indicator data $\{i(\mathbf{u}_\alpha; z_k), \alpha = 1, \ldots, n\}$ at each threshold value z_k, there is only one IK system to solve providing IK weights $\lambda_\alpha(\mathbf{u})$ that are not any more function of the threshold z_k.

Under median IK, the ccdf (III.2) takes the simpler expression:

$$F(\mathbf{u}; z|(n)) = \sum_{\alpha=1}^{n} \lambda_\alpha(\mathbf{u}) \cdot i(\mathbf{u}_\alpha; z) \tag{III.5}$$

The simplification brought by median IK comes though at a substantial cost—that of losing the flexibility of providing different indicator variogram models for different grade classes; for example, high-grade (rich) versus low-grade (poor) mineralization.

Block IK

Because block kriging can provide directly an estimate of the (linear) average grade of a block of any size $|v|$, one may think naively that block indicator kriging could also provide directly the SMU support ccdf defined as:

$$F_v(\mathbf{u}; z|(n)) = \text{Prob}\{Z_v(\mathbf{u}) \leq z|(n)\}$$

where $Z_v(\mathbf{u})$ is the RV modeling our uncertainty regarding the average grade $z_v(\mathbf{u})$ defined over the SMU $v(\mathbf{u})$ centered at location $\mathbf{u}$.

Unfortunately, block IK would provide an estimate of the average (or composite) of all point ccdfs $F(\mathbf{u}'; z|(n))$ when $\mathbf{u}'$ takes all possible positions within the block v (or V). This is *not* the sought after block ccdf $F_v(\mathbf{u}; z|(n))$. More precisely, block IK over a support $v(\mathbf{u})$ provides the composite point support ccdf:

$$F_{\cdot/v}(\mathbf{u}; z|(n)) = \frac{1}{|v|} \int_{v(\mathbf{u})} F(\mathbf{u}'; z|(n)) d\mathbf{u}' \neq F_v(\mathbf{u}; z|(n)) \tag{III.6}$$

Such composite point ccdf, when applied to a large panel V instead of a single SMU $v(\mathbf{u})$, is denoted as $F_{\cdot/V}(z|(n))$ and represents the probability that a point $\mathbf{u}'$ taken anywhere within panel V has grade no greater than the threshold value z; this is also an estimate of the proportion of points within V with grade $z(\mathbf{u}') \leq z$.

III.2 Volume-Variance Correction

At this point, indicator kriging can provide a set of point support ccdfs $\{F(\mathbf{u}_j; z|(n)),\ \mathbf{u}_j \in V\}$ over a grid of nodes $\mathbf{u}_j$ discretizing any domain or panel V. These ccdfs must be corrected for the support effect, starting with a decrease of variance going from the point to the SMU volume support.

Ideally, one would like to be able to derive the variance $Var\{Z_v(\mathbf{u})|(n)\}$ associated with each specific SMU ccdf $F_v(\mathbf{u}; z|(n))$. Note that this variance is conditional to the relation between the data retained and the specific SMU $v(\mathbf{u})$ considered; it thus varies from one SMU $v(\mathbf{u})$ to another $v(\mathbf{u}')$. Calculation of such conditional variance requires knowledge of the multivariate spatial distribution of all point RVs $Z(\mathbf{u})$ constituting the panel V, that multivariate distribution that is precisely implicit to all stochastic simulation algorithms. The only statistic that can be calculated from the sole variogram model $2\gamma(\mathbf{h})$ is the nonconditional variance of $Z_v(\mathbf{u})$ over the panel V, also called "dispersion variance" and denoted as $D^2(v/V)$ [see Journel and Huijbregts, 1987, p. 61; Isaaks and Srivastava, 1989, p. 476; Deutsch, 2002]:

$$\frac{1}{|V|} \int_V Var\{Z_v(\mathbf{u})|(n)\} d\mathbf{u} \simeq D^2(v/V) \tag{III.7}$$

The previous averaging over all possible SMU locations $\mathbf{u}$ within a large enough domain V amounts to a deconditioning, resulting in the stationary (location $\mathbf{u}$-independent) dispersion variance $D^2(v/V)$. This dispersion variance is constant within the zone/domain characterized by the point support semivariogram model $\gamma(\mathbf{h})$, and is calculated as (Journel and Huijbregts, 1978, p. 67):

$$D^2(v/V) = \bar{\gamma}(V,V) - \bar{\gamma}(v,v) \tag{III.8}$$

with:

$$\bar{\gamma}(v,v) = \frac{1}{|v|^2}\int_v d\mathbf{u} \int_v \gamma(\mathbf{u}-\mathbf{u}')d\mathbf{u}' \simeq \frac{1}{n_v^2}\sum_{i=1}^{n_v}\sum_{j=1}^{n_v}\gamma(\mathbf{u}_i-\mathbf{u}_j)$$

being the average semivariogram value between any two points $\mathbf{u}_i$ and $\mathbf{u}_j$ discretizing the SMU v, and similarly:

$$\bar{\gamma}(V,V) = \frac{1}{|V|^2}\int_V d\mathbf{u} \int_V \gamma(\mathbf{u}-\mathbf{u}')d\mathbf{u}'$$

The VVC consists of approximating the ratio of point-to-SMU support conditional variances to the known ratio of the corresponding dispersion variances (see Parker, 1980; Isaaks and Srivastava, 1989, p. 479):

$$f = \frac{Var\{Z_v(\mathbf{u})|(n)\}}{Var\{Z(\mathbf{u})|(n)\}} \simeq \frac{D^2(v/V)}{D^2(\cdot/V)} = 1 - \frac{\bar{\gamma}(v,v)}{\bar{\gamma}(V,V)} \tag{III.9}$$

where $D^2(\cdot/V) = \bar{\gamma}(V,V)$ is the dispersion variance of the point grades within the panel V.

Note that in the above expression, the sill or any factor multiplying the semivariogram model $\gamma(\mathbf{h})$ would cancel out; thus, if that model has been standardized to the data variance within panel V, that is, to $\bar{\gamma}(V,V)$, then expression (III.9) is rewritten in terms of that standardized semivariogram $\gamma_S(\mathbf{h})$:

$$f = \frac{Var\{Z_v(\mathbf{u})|(n)\}}{Var\{Z(\mathbf{u})|(n)\}} \simeq 1 - \bar{\gamma}_S(v,v) \tag{III.10}$$

hence:

$$Var\{Z_v(\mathbf{u})|(n)\} \simeq f \cdot Var\{Z(\mathbf{u})|(n)\}$$

The support and information effects on the SMU conditional variance $Var\{Z_v(\mathbf{u})|(n)\}$ are split in two factors: the support effect is accounted for by the stationary factor f approximated by the ratio of dispersion variances, the information effect is accounted for by the nonstationary (location $\mathbf{u}$-dependent) point variance $Var\{Z(\mathbf{u})|(n)\}$.

VVC before or after within-panel compositing

The volume-variance correction, as expressed in relation (III.9), arises from a very common paradigm of estimation, one that consists of estimating ratios of unknown values rather than estimating each of the unknown values themselves. Indeed, ratios are usually more stable, and hence more conducive to the stationarity (permanence) hypothesis underlying most estimation endeavors.

The permanence hypothesis underlying expression (III.10) relates to the ratio of variances of the two location-specific ccdfs $F_v(\mathbf{u}; z|(n))$ and $F(\mathbf{u}; z|(n))$. A similar permanence hypothesis could be applied to the ratio of the variances of the two composite ccdfs averaged over the panel V; that is, the two ccdfs:

$$F_{\cdot/V}(z|(n)) = \frac{1}{N}\sum_{j=1}^{N} F(\mathbf{u}_j; z|(n)) \tag{III.11}$$

$$F_{v/V}(z|(n)) = \frac{1}{N}\sum_{j=1}^{N} F_v(\mathbf{u}_j; z|(n))$$

where the panel $V = Nv$ is discretized by N SMUs centered at locations $\{\mathbf{u}_j, j = 1, \ldots, N\}$. In the first expression (III.11), the SMU size is actually a point.

Let $Var\{\cdot/V|(n)\}$ be the variance of the point support composite ccdf $F_{\cdot/V}(z|(n))$ within panel V, and $Var\{v/V|(n)\}$ be the variance of the SMU composite ccdf $F_{v/V}(z|(n))$; the permanence hypothesis would lead to the following variance ratio approximation similar to that expressed in (III.10):

$$f = \frac{Var\{v/V|(n)\}}{Var\{\cdot/V|(n)\}} \simeq 1 - \bar{\gamma}_S(v, v) \tag{III.12}$$

The volume-variance factor f defined in relation (III.10) is said to be "before compositing", whereas the factor f defined in relation (III.12) is said to be "after compositing". Compositing here relates to the averaging (III.11) of ccdfs within the domain or panel V.

Which is best?
Dispersion variances, as expressed in relation (III.7), are in fact composite variances averaged over a domain V large enough to remove the (nonstationary) influence of specific data values. On this ground, one could argue that the permanence hypothesis should apply to the ratio of variances of composite ccdfs as in (III.12).

However, composite variances, as defined in relation (III.7), are not the variances of the composite ccdfs defined in relations (III.11). Therefore the previous argument is not bulletproof. In addition, there is one advantage to applying VVC before compositing: the correction (III.10) would provide the variance of the SMU ccdf $F_v(\mathbf{u}; z|(n))$ at any specific location $\mathbf{u}$ within panel V, instead of the variance of a single composite SMU ccdf $F_{v/V}(z|(n))$. One could counterargue that the permanence hypothesis is not good enough to be applied to each individual SMU ccdf, and it is safer to restrict it to the composite SMU ccdf.

Our position is that it is time to give some rest to the VVC approach under all its variants, and replace it with the simulation approach, which is more versatile and provides greater internal consistency.

If hard-pressed to take a position between VVC before or after compositing, we would suggest trying both on a few test panels V:

- If the results are similar, we would prefer the more versatile option (III.10); that is, a correction applied before compositing the ccdfs.
- If the results are significantly different, we would stay away from the VVC approach as a whole: the ratios of variances appear not to be stable enough to justify a VVC.

III.3 Permanence of Distribution

At this point we have available:

- a grid of point support ccdfs $\{F(\mathbf{u}_j; z|(n)), \mathbf{u}_j \in V\}$, and
- the variances of the SMU support ccdfs, $Var\{Z_v(\mathbf{u}_j)|(n)\}$ as given by the VVC before compositing in relation (III.10), or the variance of the composite SMU ccdf, $Var\{v/V|(n)\}$ as given by the VVC after compositing in relation (III.12).

The final goal is to get an estimate (actually a model) for either the SMU ccdfs $\{F_v(\mathbf{u}_j; z|(n)), \mathbf{u}_j \in V\}$ or for the composite SMU ccdf $F_{v/V}(z|(n))$. We thus need to transform the known (IK-derived) point support ccdfs, either $\{F(\mathbf{u}_j; z|(n)), \mathbf{u}_j \in V\}$ or the composite ccdf $F_{./V}(z|(n))$, to match the smaller conditional variances of the SMU support. To this goal, a permanence of distribution type is called for.

Ideally, one would like to have access to a histogram of SMU support average grades $z_v(\mathbf{u})$, say the marginal cdf:

$$F_v(z) = \text{Prob}\{Z_v(\mathbf{u}) \leq z\}$$

Such a histogram would provide precious information about how averaging over a support v affects (reduces) the tails of the histogram of the quasi-point support grades $z(\mathbf{u})$. Instead, the only information typically available at the time of prediction is:

- the data support marginal cdf $F(z) = \text{Prob}\{Z(\mathbf{u}) \leq z\}$, and
- the empirical observation that, if the distribution $F(z)$ of core metal grades is approximately lognormal, the distribution $F_v(z)$ over "not too large" block v is also approximately lognormal (David, 1977; Journel and Huijbregts, 1978; Rendu, 1979).

This permanence of lognormality was at the origin of many support correction algorithms: the block ccdf $F_v(\mathbf{u}; z|(n))$ was assumed lognormal with mean either obtained directly by block kriging or identified to the mean of the central point ccdf $F(\mathbf{u}; z|(n))$; there remained to determine the block ccdf variance, this was done using various approximations, the most common being the VVC factor defined in relation (III.10) [see also David, 1988].

However, assuming that all block ccdfs are lognormal, whatever the block size $|v|$ and whatever the conditioning data (n), is a very restrictive hypothesis. Isaaks and Srivastava (1989, p. 472) proposed instead to apply the lognormal analytical correction (see following) to the known point support ccdf $F(\mathbf{u}; z|(n))$ or composite ccdf $F_{\cdot/V}(z|(n))$. The original point support ccdf need not be lognormal, nor is the resulting corrected SMU support ccdf lognormal, yet at the limit if the point support ccdf is indeed lognormal, that analytical correction would yield a lognormal SMU support ccdf, thus respecting the empirically observed permanence of lognormality. For convenience, we have retained this "indirect lognormal correction" because it is implemented in the GSLIB program `postik` (Deutsch and Journel, 1998, p. 236).

In the general case, the indirect lognormal correction does not result in a lognormal block ccdf; instead, the shape of the block ccdf is gradually modified from that of the point ccdf, becoming less skewed (more symmetric with shorter tails) as the block size $|v|$ increases, and correspondingly, as the variance ratio f decreases.

The indirect lognormal correction (recall)

Consider two lognormal cdfs $F(z)$ and $F_v(z)$ with the same mean and different variances $\sigma^2 > \sigma_v^2$, the point support variance σ^2 being necessarily greater than the block support variance σ_v^2. Define the variance correction ratio as:

$$f = \sigma_v^2/\sigma^2 \quad \in [0, 1]$$

The quantile function $q(p)$ is defined as the inverse of the cdf $F(z)$; that is, the z-value that has probability p not to be exceeded:

$$q(p) = F^{-1}(p) = z \text{ value s.t. } F(z) = p$$

Note that the mean m and variance σ^2 of the cdf $F(z)$ are expressed in terms of either the cdf $F(z)$ or the quantile function $q(p)$:

$$m = E\{Z\} = \int_{-\infty}^{+\infty} z dF(z) = \int_0^1 q(p) dp$$

$$\sigma^2 = E\{Z^2\} - m^2$$

with:

$$E\{Z^2\} = \int_{-\infty}^{+\infty} z^2 dF(z) = \int_0^1 q^2(p) dp$$

Quantile function integrals are particularly simple to evaluate. It is therefore convenient to express change of support algorithms in terms of quantile functions if means and variances are to be controlled. The quantile transform going from a lognormal distribution with parameters (m, σ^2) to another lognormal distribution with parameters (m, σ_v^2) is written (Isaaks and Srivastava, 1989, p. 473–475):

$$\begin{aligned} q_v^{(0)}(p) &= a[q(p)]^b \\ \text{or} \quad \ln q_v^{(0)}(p) &= \ln a + b \ln q(p) \end{aligned} \tag{III.13}$$

where $q_v^{(0)}(p)$ is the quantile function of a lognormal distribution with parameters (m, σ_v^2), and $q(p)$ is the quantile function of another lognormal distribution with parameters (m, σ^2). The coefficients a, b are given by the expressions:

$$a = \frac{m}{\sqrt{fCV^2+1}} \left[\frac{\sqrt{CV^2+1}}{m}\right]^b \qquad b = \sqrt{\frac{\ln(fCV^2+1)}{\ln(CV^2+1)}}$$

where:

$$CV = \sigma/m, \qquad f = \sigma_v^2/\sigma^2 \in [0,1]$$

Starting from a lognormal quantile function $q(p)$, the transform (III.13) would restitute exactly the desired mean m and reduced variance $\sigma_v^2 = f\sigma^2$. However, if applied to a nonlognormal point quantile function $q(p)$, this transform would not restitute the common mean m. Hence a mean identification is added to the transform (III.13):

$$q_v(p) = \frac{m}{m^{(0)}} q_v^{(0)}(p) \tag{III.14}$$

with $m^{(0)} = \int_0^1 q_v^{(0)}(p)dp$ being the mean associated to the quantile function $q_v^{(0)}(p)$ defined in relation (III.13), and $m = E\{Z\} = E\{Z_v\}$ being the target mean.

In practice, starting from the point support quantile function $q(p)$, which need not be strictly lognormal, application of the transform (III.13) yields the quantile function $q_v^{(0)}(p)$. The corresponding mean value $m^{(0)}$ is calculated and used in the final transform expression (III.14). Provided the variance correction is not too large ($f > 0.7$), the result is a block support quantile function $q_v(p)$ that features the target mean m, a variance close but not identical to the target variance $\sigma_v^2 = f\sigma^2$, and a distribution shape close to that of the original point support distribution.

The algorithm yields "reasonably looking" block cdfs if the original point cdfs are close to being lognormal. As the correction ratio f decreases, the variance correction is more important and the block cdf shape departs more from that of the point cdf. The fact that the transform (III.14) does not yield back exactly the target block variance $\sigma_v^2 = f\sigma^2$ is not seen as a major impediment, because this block variance is itself an approximation resulting from VVC. An alternative is to modify ("fudge") the correction ratio f until the variance of the distribution III.14 matches the target variance σ_v^2.

Alternative algorithms

The actual effect of the indirect lognormal correction (III.13 and III.14) and the reason why "it works" is the gradual symmetrization of the block cdf, $F_v(z)$ or $F_v(\mathbf{u}; z|(n))$, around the mean of the point cdf $F(z)$ or $F(\mathbf{u}; z|(n))$. Any other algorithm based on some form of permanence of distribution-type hypothesis that would achieve the same gradual symmetrization would be acceptable, independent of its theoretical pedigree. A theoretical pedigree not backed by data (e.g., actual block support average grade values) is of no great practical value, particularly if the "theory" is poorly understood.

The different algorithms for modeling a block cdf (or ccdf) from a point support cdf differ by the degree of symmetrization brought to the initial point support cdf for a given variance correction ratio f.

At one end of the spectrum, there is the affine correction, which keeps the point cdf shape as is; that is, brings no symmetrization (see Journel and Huijbregts, 1978, p. 471; or Isaaks and Srivastava, 1989, p. 471). Unless the block size $|v|$ is very small and there is little to no variability of point grades within v (extreme small scale continuity), we recommend not using this extreme permanence hypothesis.

At the other end of the spectrum, when the block size $|v|$ is large and the entropy of the point grade spatial distribution within V is large, the block cdf would tend to be normal, although with zero probability for negative values. In such a case, a pseudo-Gaussian symmetric distribution for the block cdf would be appropriate. This is implicitly what the discrete Gaussian model underlying disjunctive kriging delivers (Journel and Huijbregts, 1978, p. 472). The symmetrization provided by the discrete Gaussian model is more drastic than that provided by the indirect lognormal correction (for the same variance correction ratio f) and may explain the "good" results obtained with large SMU v in the absence of large-scale spatial structures; that is, in the presence of high spatial entropy.

The distribution type of block cdfs depends on the spatial entropy; that is, the degree of organization of the mineralization, both within the SMU v and within the panel V. That spatial entropy characteristic is not reflected in the variogram (a mere two-point statistic), and hence, in the absence of additional data it is difficult to decide a priori which is the "best" block distribution model. Relevant additional data could be a histogram of actual block data, or an exhaustive training image at the point support level deemed representative of the spatial distribution of point grades within a panel V. The latter (training images) is precisely the information used by modern simulation algorithms to derive simulated images of the point grade spatial distributions, which could then be averaged into SMU support grade distributions (see Caers and Journel, 1998; and Strebelle, 2002).

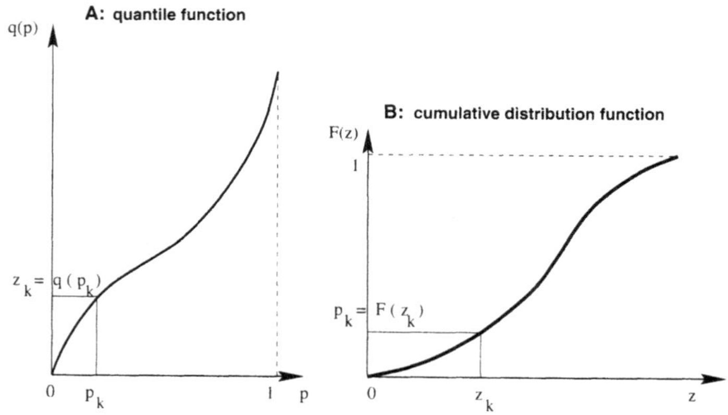

Figure III.1: Going from a quantile function $q(p)$ (**A**) to a cdf $F(z)$ (**B**). The inversion $F(z) = q^{-1}(z)$ or $q(p) = F^{-1}(p)$ amounts to simply inverting the abscissa and ordinate axes.

III.4 Estimates of Recovery Functions

Using the volume-variance correction (VVC) and an indirect lognormal correction (or any alternative hypothesis for permanence of distribution), we have now access to a quantile function representation of:

- either every specific SMU ccdf centered at location $\mathbf{u}_j$:

$$F_v(\mathbf{u}_j; z|(n)) = \text{Prob}\{Z_v(\mathbf{u}_j) \leq z|(n)\}, \quad \forall \mathbf{u}_j \in V$$

  this is the output of VVC before compositing within panel V,

- or the SMU composite ccdf:

$$F_{v/V}(z|(n)) = \frac{1}{N} \sum_{j=1}^{N} F_v(\mathbf{u}_j; z|(n))$$

  this is the output of VVC after compositing within panel V.

From quantile functions to cdfs

If the previous ccdfs are obtained as quantile functions, inversion into the desired ccdf values is trivially done through a graphical transform (see Fig. III.1).

More precisely, consider the quantile function $q(p) = F^{-1}(p)$ corresponding to any cdf $F(z)$. The function $q(p)$ is plotted versus discrete values of p—typically decile values corresponding to p varying by 0.1 increments or centile values corresponding to p varying by 0.01 increments (see Fig. III.1A). Inversion of this graph corresponds simply to inverting the abscissa and ordinate axes, reading $F(z) = p$ from $z = q(p)$ [see Fig. III.1B]. Proper interpolation between the original $\{p_k, q(p_k)\}$ points discretizing the quantile function $q(p)$ is required (see Deutsch and Journel, 1998, p. 134). Such interpolation/extrapolation is routinely performed in various programs of the GSLIB software, in particular the program `postik` used to retrieve recovery functions from IK-derived ccdfs. Other implementations are also possible (e.g., Barnes, 1999).

Extracting recovery functions
Consider, for example, the SMU composite ccdf $F_{v/V}(z|(n))$, interpreted as the probability for the average grade $Z_v(\mathbf{u})$ of a SMU v taken anywhere within V to be no greater than the cutoff grade z. The following recovery functions can be retrieved from that composite ccdf:

- tonnage recovery within V:

$$t_{v/V}(z) = N\left[1 - F_{v/V}(z|(n))\right] \tag{III.15}$$

- quantity of metal recovered:

$$q_{v/V}(z) = N \int_z^{\infty} y dF_{v/V}(y|(n))$$

- mean ore grade:

$$m_{v/V}(z) = q_{v/V}(z)/t_{v/V}(z)$$

  this is the truncated mean (above z) of the ccdf $F_{v/V}(z|(n))$,

- profit: a known (deterministic) function of the previous tonnage and quantity of metal recovered. That function need not be specified here.

III.5 Summary

We chose the VVC algorithm to provide a yardstick for evaluating the performance of the proposed simulation approach. The VVC algorithm is reasonably well understood by practitioners aware of the support effect on ore reserves recovery. In the VVC approach, probability distributions of unsampled point support grades are built from the point support data; this is typically done through some form of indicator kriging. These point support distributions are then corrected for the volume support of SMUs by:

- decreasing the point support variance using a permanence hypothesis of ratios of variance (from point to SMU)—this is where VVC specifically applies—and

- assuming some distribution type (shape) for the SMU support histograms.

If the VVC approach does account for the support effect, it *does not* account for the information effect; that is, for the fact that SMUs are selected on the basis of their future estimated values. Because of smoothing, these future estimated values are likely to have variances and distribution shapes different from the true SMU values; yet it is the distribution of the latter that the VVC approach aims at modeling. This important limitation of the VVC approach is fully stressed in Chapter VI, where VVC is applied.

Chapter IV

The Reference Data Sets

To provide a demonstration of the methodology proposed for prediction of recoverable reserves, one would like to have access to a large, exhaustive data set including the sample data used for prediction and those used for selection, as well as the actual grades of all SMUs considered. Unfortunately, such information related to a real mining operation is rarely (if ever) available. Only very large operations with computerized track records of all drilling, sampling, and actual recovery characteristics could provide exhaustive data sets; even there, true SMU grades always remain unknown.

The next best alternative to a real mining data set is to consider a "natural" data set that has not been synthesized by a simulation algorithm that could bias the demonstration proposed. Not uncommon, such recursive bias would consist of using a reference data set generated by readily available multivariate Gaussian algorithms to test methodologies that rely precisely on properties of the Gaussian distribution. Topographic elevation data sets are widely available, they are very accurate in the sense that biases and artefact structures brought by the measurement/interpolation process are minimum, they can be very extensive approaching the exhaustive character we need, they are available on a regular grid that is convenient for averaging, last they are in the public domain, and hence can be released to anyone wishing to repeat the calculations proposed.

We decided to retain for reference point support data set the digital elevation model taken from the United States Geological Survey (USGS) quadrangle Ely West, near the town of Ely in eastern Nevada.

The Ely data set defined on a point support is presented in Section IV.1. Using these exhaustive reference data, the recovery functions under perfect selection are calculated in Section IV.2. In practice, such perfect selection cannot be achieved, because actual selection is based on SMU estimates not on the true SMU values. Section IV.3 presents an example of simulation of future poorer-quality selection data (blastholes). From these selection data, future estimates of SMU average grades can be derived, allowing determination of actual recovery under imperfect selection. In Section IV.3, once a SMU has been selected as ore on the basis of its future estimated value, the grade recovered is the true SMU grade as obtained from the exhaustive reference data set. This exercise allows an appreciation of the effect of imperfect selection on recovery and profit. The last section introduces a second reference data set, which better approximates the complex mineralizations found in precious metal deposits. This second data set is used for checking the robustness of the algorithms proposed for recovery prediction.

IV.1 The Ely Reference Data Set

The original Ely West elevation data did not display the typical highly skewed histogram of mineral grades. We decided to transform these original elevation data to approximate the more skewed histogram of the well published Walker Lake data set (see Isaaks and Srivastava, 1989, p. 72, and Fig. IV.1B). The transform used is a quantile transform matching the p-quantile of the original elevation data histogram to the corresponding p-quantile of the Walker Lake data histogram (see the program **trans** in Deutsch and Journel, 1998, p. 227).

All statistics shown in Figures IV.1 and IV.2 refer to the grayscale pixel map of Figure IV.1A, which represents from now on our first true and exhaustive reference data set, called the Ely reference data set.

The Ely reference data set consists of 286 × 286 quasi-point values with mean 265.34 grade units and a large coefficient of variation 1.82 (see histograms of Figs. IV.1B–D). There is a significant proportion 5.83% of zero grades (4771 values). The unit for the grade values is not important: It could be seen as *ppm* or g/t, or $0.01g/t$ corresponding to a gold mine with average grade $2.65g/t$. What is important is the long-tailed asymmetric histogram, which is neither lognormal nor exponential. Note the maximum value 6740, one order of magnitude larger than the mean 265 and two orders of magnitude larger than the median value 57.

Figure IV.1A displays various arc-like structures of high grades elongated approximately along NS. The reasonably repetitive character of those structures would make a model decision of stationarity plausible. This is the very reason we chose to retain the original Ely West data set instead of the better published Walker Lake data set. The 5.8% of zero values tend to cluster together in continuous low flats (see Fig. IV.1C).

The directions of major and minor spatial continuity are identified from the normal score semivariogram map of Figure IV.2A. The original z-reference data were transformed into normal scores y-data with standard normal distribution using the GSLIB program **nscore** (Deutsch and Journel, 1998, p. 223). The direction of maximum continuity was taken to be N10°E, whereas the direction of minimum continuity is perpendicular at N80°W. Figures IV.2B and IV.2C give the exhaustive reference semivariograms along these two principal directions for the normal score y-data and the original z-data, together with the fit from the two-dimensional semivariogram models adopted. The reference semivariograms were standardized to their respective variances, 1 for $\gamma_Y(\mathbf{h})$, $(483)^2$ for $\gamma_Z(\mathbf{h})$.

The two-dimensional anisotropic variogram models are:

$$\gamma_Y(\mathbf{h}) = 0.65\text{Gauss}\sqrt{\left(\frac{h_1}{60}\right)^2 + \left(\frac{h_2}{35}\right)^2} + 0.35\text{Exp}\sqrt{\left(\frac{h_1}{230}\right)^2 + \left(\frac{h_2}{37}\right)^2} \quad \text{(IV.1)}$$

$$\gamma_Z(\mathbf{h}) = 0.65\text{Exp}\sqrt{\left(\frac{h_1}{70}\right)^2 + \left(\frac{h_2}{35}\right)^2} + 0.35\text{Exp}\sqrt{\left(\frac{h_1}{75}\right)^2 + \left(\frac{h_2}{40}\right)^2} \quad \text{(IV.2)}$$

where $\mathbf{h} = (h_1, h_2)$ is the separation vector with components h_1, h_2 along the principal directions of anisotropy N10°E and N80°W, respectively; Gauss$(\cdot)$ denotes a Gaussian semivariogram component with major practical range 60 units and minor practical range 35 units; and similarly for the exponential component Exp$(\cdot)$.

The lack of nugget effect is characteristic of the short scale continuity of topographic elevation data and is not representative of mineral grades. Thus, this reference data set would be more akin to BTU (British thermal unit) grades in a coal seam than precious metal grades in a Carlin-type deposit. In retrospect, we should have considered elevation data over a more rugged terrain. This will be corrected in the generation of the second reference data set in Section IV.4.

Note the hole effect (dip) of the normal score reference semivariogram along direction N80°W at around $h_2 \simeq 70$: this corresponds to the pseudo-periodicity of the alignment of high-valued ridges; this hole effect was not modeled in this study.

Checking bivariate normality

Because we will later consider Gaussian-based stochastic simulation algorithms, we checked the bivariate Gaussian assumption for the normal score y-data. These y-data are already univariate standard normal per the normal score transform, but this transform does not guarantee that the bivariate distribution of any pair of random variables $\{Y(\mathbf{u}), Y(\mathbf{u}+\mathbf{h})\}$ be binormal. The check is performed by computing a series of indicator semivariograms of the y-data and comparing them to their binormal theoretical expression deduced from the model $\gamma_Y(\mathbf{h})$ given in relation (IV.1) [see Goovaerts, 1997, p. 271]. The GSLIB program `bigaus` was used to this purpose.

Figure IV.3 gives the reference indicator semivariograms along the direction of maximum continuity N10°E and their fit by the theoretical binormal models for the four p-quantile threshold values $\{z(p), p = 0.05, 0.5, 0.95, 0.99\}$. Figure IV.4 gives the corresponding results along the direction N80°W of minimum continuity. The deviations observed between the experimental curves and the binormal expression would be considered acceptable in the practice of multivariate Gaussian simulation (compare to Fig. 7.6 on page 274 of Goovaerts, 1997). Note that the reference N10°E indicator semivariogram at low threshold ($p = 0.05$) is more continuous than (below) the theoretical binormal curve: this is explained by the extreme spatial continuity of flat zero beds. Conversely, the high-threshold ($p = 0.95, 0.99$) indicator semivariograms indicate that the reference high z-values are more disconnected in space than modeled by a binormal distribution.

We will hereafter accept the multiGaussian model for the normal score random function $\{Y(\mathbf{u}), \mathbf{u} \in D\}$.

Vein-controlled gold mineralizations typically show more continuity at high-threshold values than at low or median thresholds, because of structural control. Once again, this reference data set relates to topographic elevation data more akin to coal quality grades than precious metal grades. For gold deposits, a multiple indicator model coupled with simulation is usually more appropriate because of the greater flexibility offered by the multiple indicator variograms, each modeling spatial variability of a different class of grade values (Journel and Isaaks, 1984; Journel and Alabert, 1989; Dimitrakopoulos and Dagbert, 1993).

Reference SMU data

The previous reference "point" support z-data are averaged into block support SMU values. The two SMU sizes here considered are 11×11 and 22×22 (see Figs. IV.5C–E). The dimension 286×286 of the reference point support field is such that this averaging results into an exact integer number of SMUs: $N = 26 \times 26 = 676$ SMUs of size 11×11, and $N = 13 \times 13 = 169$ SMUs of size 22×22.

The arithmetic (linear) averaging does not affect the mean value (265.34) but reduces the coefficient of variation, essentially through shortening of the histogram tail (see Figs. IV.5D–F). The proportion of low values (<100), however, does not change significantly with the block averaging.

The reference point support pixel plot and histogram have been repeated in Figures IV.5A and IV.5B for easier comparison. Note the gradual loss of resolution caused by averaging. Also, averaging increases the median value from 57 to 76 (11×11 SMUs) and 115 (22×22 SMUs), yet the SMU histograms remain clearly asymmetric.

Figures IV.6A and IV.6B give the quantile-quantile plots of the SMU support grade distributions versus the point support grade distribution: these are plots of the SMU p-quantile values versus the point p-quantile grades (same p value). The relations between the point p-quantiles and SMU p-quantiles are non-linear, which implies that the two distribution shapes cannot be made similar by a mere affine correction (see Isaaks and Srivastava, 1989, p. 471). In short, point and block grade distributions have different shapes.

The semivariograms of the reference point and SMU values along the two principal directions are plotted on Figures IV.6C and IV.6D. These semivariograms were all standardized to the variance $(483)^2$ of the reference point support grades. As expected, the block-support semivariograms present a lower sill (spatial variance) and a more continuous behavior near the origin because of within-block averaging. For a theoretical analysis of the smoothing effect caused by volume averaging, refer to Journel and Huijbregts (1978, p. 77) or Isaaks and Srivastava (1989, p. 93).

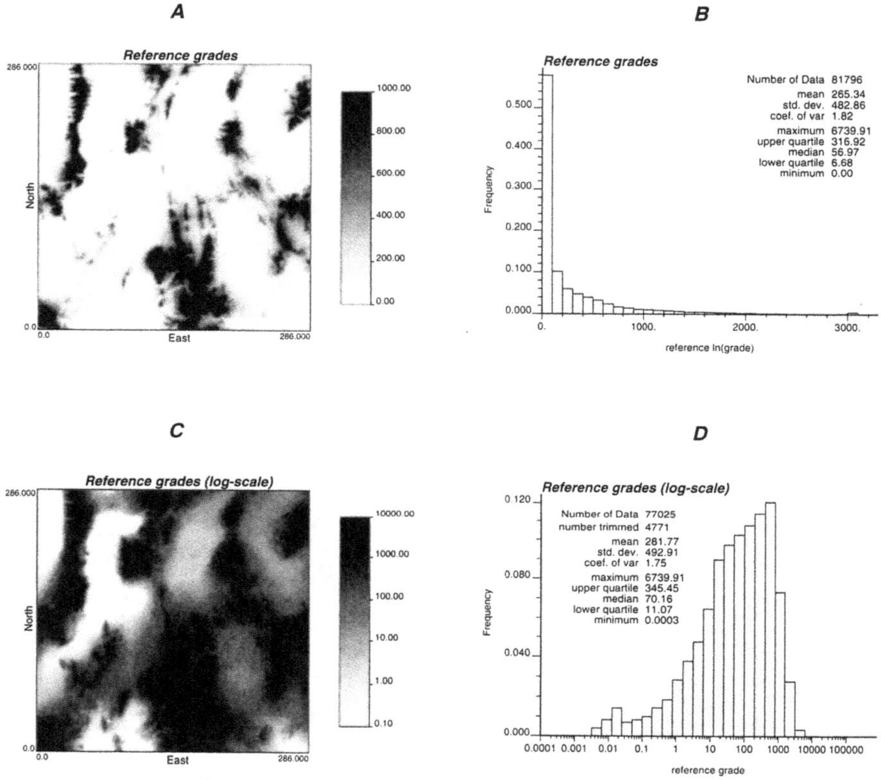

Figure IV.1: Grayscale plots (left) and histograms (right) of the Ely reference data set. **A**, **B:** arithmetic scale displaying structures of high values, **C**, **D:** logarithmic scale displaying structures of low values. The 4771 zero values are trimmed out of the histogram and statistics of **D**.

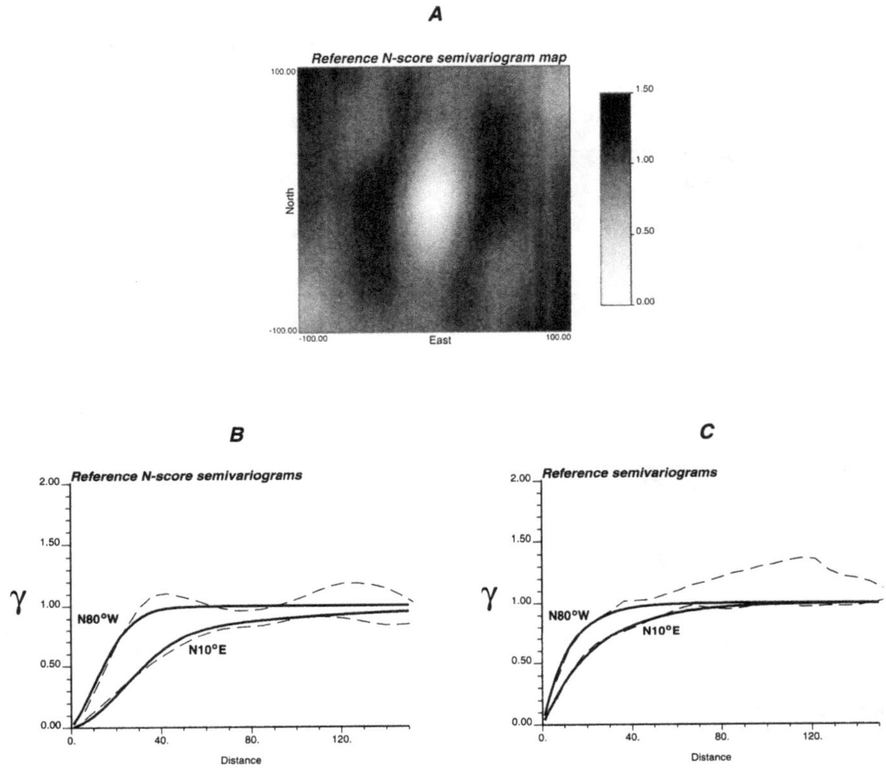

Figure IV.2: Ely reference semivariograms and model fit. **A:** normal scores semivariogram map, **B:** normal scores semivariograms in directions N10°E and N80°W, **C:** z-semivariograms in directions N10°E and N80°W; model fit is shown as solid lines.

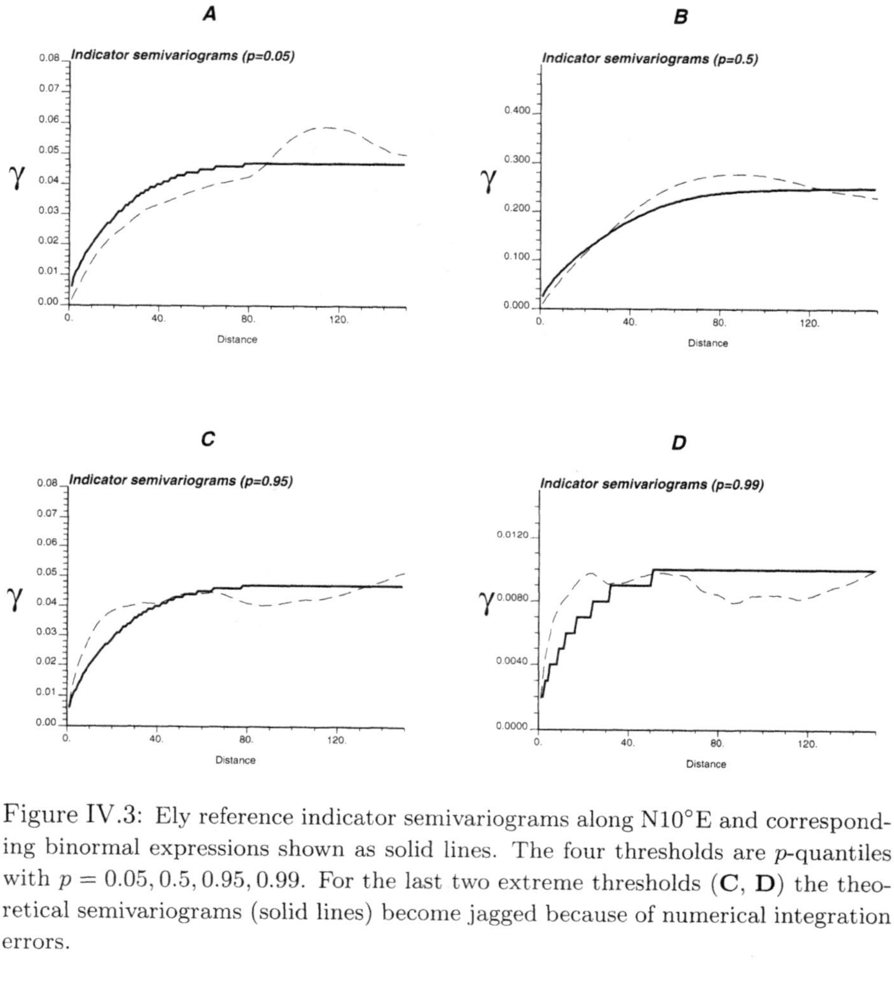

Figure IV.3: Ely reference indicator semivariograms along N10°E and corresponding binormal expressions shown as solid lines. The four thresholds are p-quantiles with $p = 0.05, 0.5, 0.95, 0.99$. For the last two extreme thresholds (**C**, **D**) the theoretical semivariograms (solid lines) become jagged because of numerical integration errors.

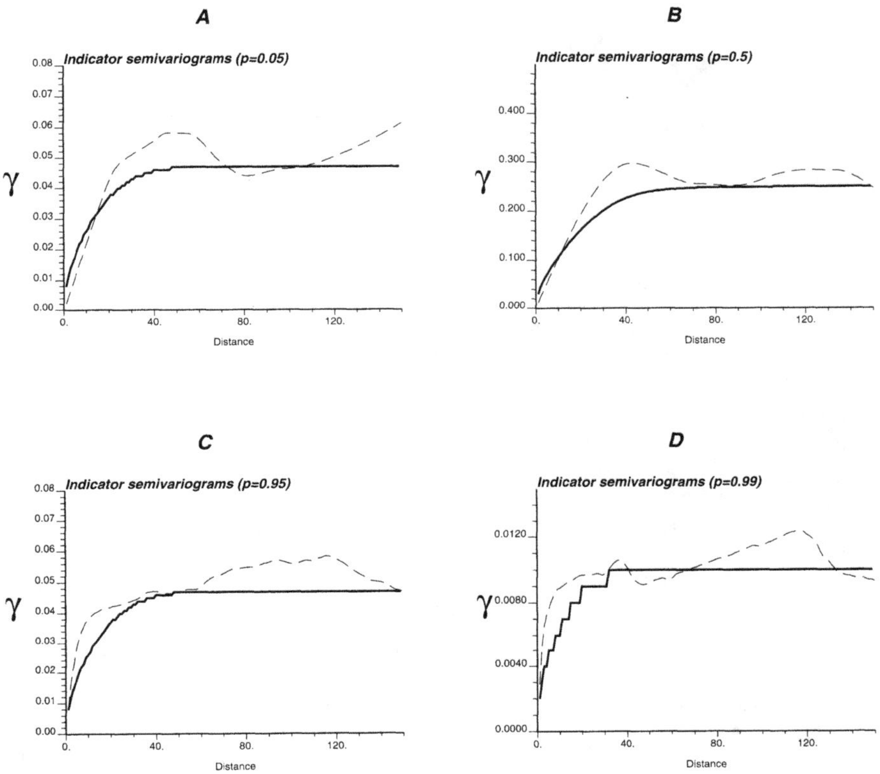

Figure IV.4: Ely reference indicator semivariograms along N80°W and corresponding binormal expressions shown as solid lines. The four thresholds are p-quantiles with $p = 0.05, 0.5, 0.95, 0.99$. For the last two extreme thresholds (**C**, **D**) the theoretical semivariograms (solid lines) become jagged because of numerical integration errors.

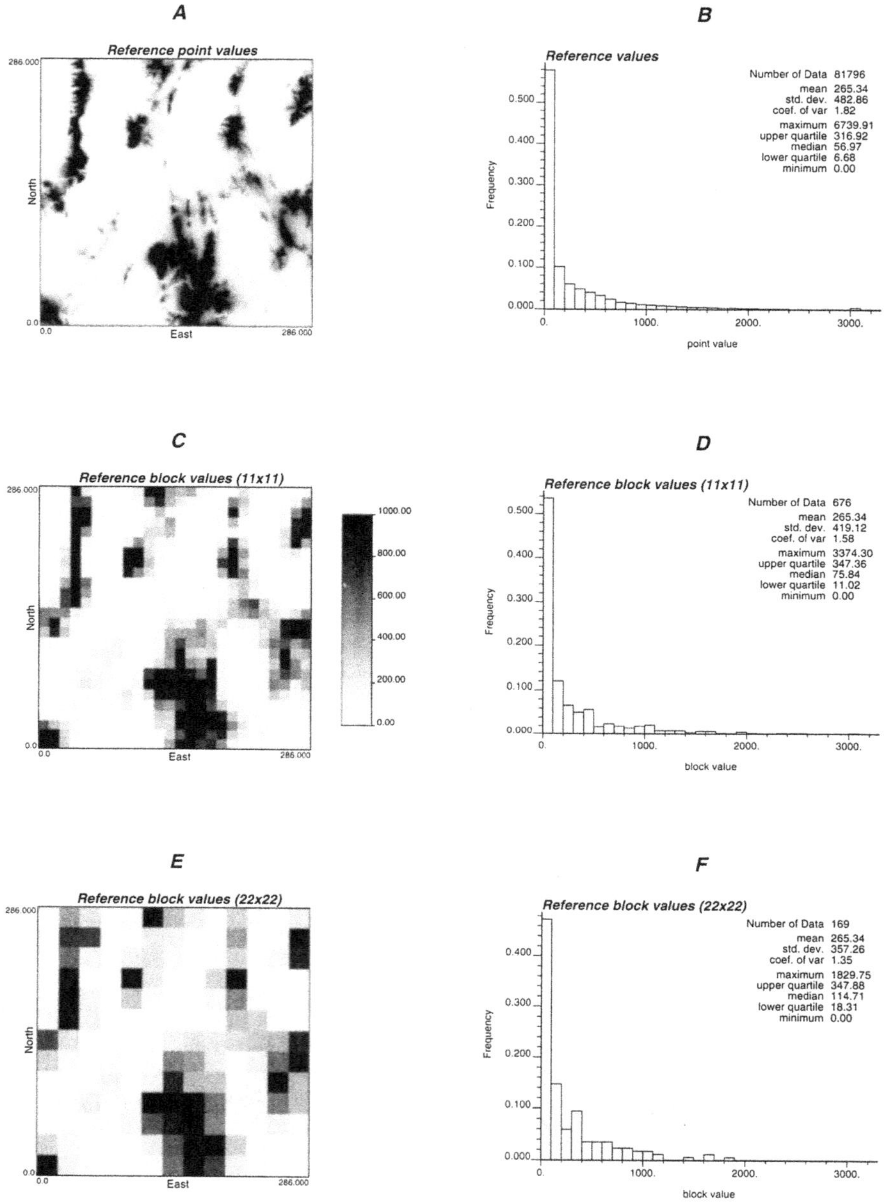

Figure IV.5: Grayscale plots (left) and histograms (right) of the Ely reference point and block support grades. **A**, **B:** point support grades and their histogram, **C**, **D:** 11 × 11 SMU grades and their histogram, **E**, **F:** 22 × 22 SMU grades and their histogram.

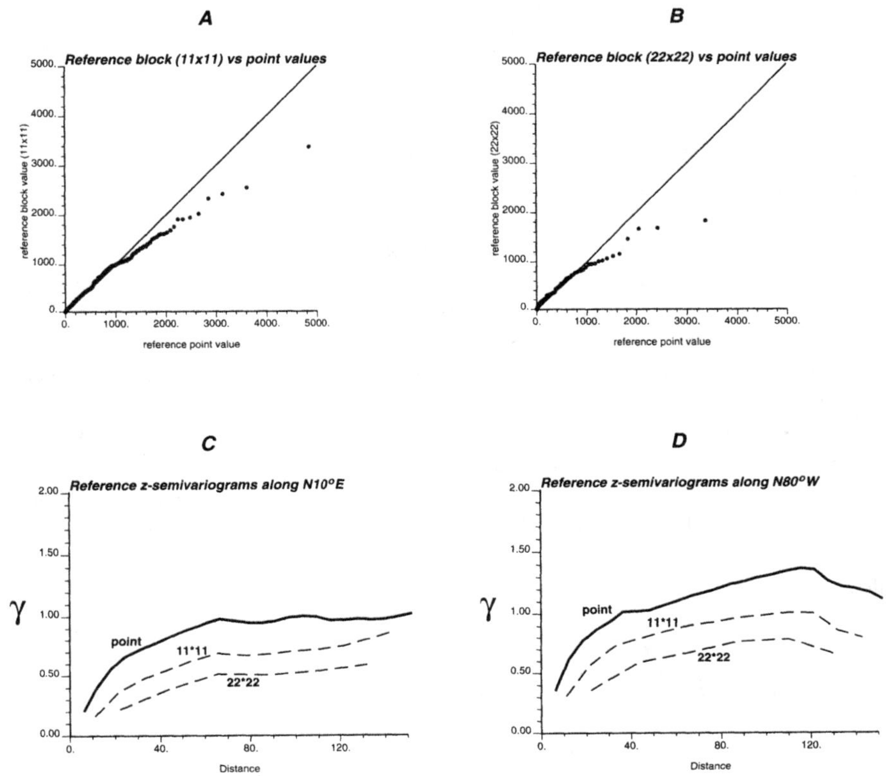

Figure IV.6: Statistics of Ely reference point and block support grades. **A**, **B:** quantile-quantile plots of block versus point distributions, **C**, **D:** semivariograms along N10°E and N80°W [all standardized to the variance $(483)^2$ of the point support grades].

IV.2 Reference Perfect Selection

Perfect selection is performed on the reference "true" SMU grades $z_v(\mathbf{u})$ [see relations (I.1) through (I.5)]. Five grade cutoffs $z = F_v^{-1}(p)$ corresponding to the following p-quantiles of the reference 11×11 SMU grade distribution $F_v(z)$ were retained consistently throughout the entire case study to follow:

$$\begin{array}{ll} p = 0.5 & z = F_v^{-1}(p) = 76, \quad |v| = 11 \times 11 \\ p = 0.6 & z = 139 \\ p = 0.7 & z = 271 \\ p = 0.8 & z = 445 \\ p = 0.9 & z = 854 \end{array} \tag{IV.3}$$

The maps of the perfect selection indicators $i_v^0(\mathbf{u}; z)$ are given in Figure IV.7 for the 11×11 SMU size and in Figure IV.8 for the 22×22 SMU size. As expected, as the cutoff z increases the number of profitable SMUs (black pixels) of a given size decreases; compare these recoverable SMU maps to the reference maps of Figure IV.5. Remember, though, that these recovery indicators correspond to perfect selection based on the true SMU grades.

Table IV.1 gives the perfect recovery figures (tons, quantity of metal and ore grade) corresponding to the five cutoff values given in (IV.3) for SMU sizes 11×11 and 22×22. Figure IV.9 gives the corresponding recovery curves as function of the cutoff grade z. Tonnage and quantity of metal are given in percent recovery of the total corresponding to zero cutoff (all SMUs are recovered). The deposit size and total unit tons are $286 \times 286 = 81796$. The mean ore grade is given in the same units as the z-cutoff grade.

The profit is expressed in million dollars ($M\$$) and is calculated as the following function of the cutoff value z:

$$p_v^0(z) = C(z) \cdot \left[q_v^0(z) - t_v^0(z) \cdot z\right] - C_0 \tag{IV.4}$$

with $C_0 = 200M\$$ for 11×11 SMU size, and $C_0 = 150M\$$ for 22×22 SMU size.

The proportionality factor $C(z)$ is the same for both SMU sizes and increases with the cutoff grade z:

$$\begin{array}{ll} z = 76 & C(z) = 20 \times 10^{-6} \\ z = 139 & C(z) = 30 \times 10^{-6} \\ z = 271 & C(z) = 40 \times 10^{-6} \\ z = 445 & C(z) = 50 \times 10^{-6} \\ z = 854 & C(z) = 55 \times 10^{-6} \end{array} \tag{IV.5}$$

As expected, the tonnage and quantity of metal recovered decrease as the cutoff increases (Figs. IV.9A and IV.9B), whereas the reverse is true for the mean ore grade recovered (Fig. IV.9C). Differences between tonnage recovered at the 11×11 and 22×22 SMU sizes are more pronounced for median cutoffs ($z = 76$ and 139), whereas this discrepancy for quantity of metal is maximum at the highest cutoff $z = 854$.

11 × 11 SMU				
Cutoff (z)	% Tonnage	% Q. of Metal	Mean Grade	Profit
76	50.0	96.6	512	157
139	40.1	92.6	612	265
271	30.0	85.2	752	272
445	20.0	71.8	952	214
854	10.1	49.1	1292	-2
22 × 22 SMU				
Cutoff (z)	% Tonnage	% Q. of Metal	Mean Grade	Profit
76	55.0	96.0	462	198
139	48.5	93.4	510	292
271	32.0	81.7	678	275
445	21.3	68.0	846	200
854	7.7	35.2	1214	-25

Table IV.1: Reference recovery values (Ely data set) for SMU sizes 11×11 and 22×22 under perfect selection. % tonnage refers to total tons $= 286 \times 286 = 81796$ unit tons (at zero cutoff), % quantity of metal refers to total $= 81796 \times 265$ unit tons $\times$ grade unit (265 is the overall mean grade), and profit is given in $M\$$.

Maximum profit ($272M\$$) for the 11×11 SMU is reached at the third cutoff grade, whereas maximum profit ($292M\$$) is reached at the second cutoff grade for the 22×22 SMU size. Note the high penalization of the few SMUs recovered at the highest cutoff $z = 854$, which leads to a significant negative profit figure ($-25M\$$) for the 22×22 SMU size, and a lesser negative profit ($-2M\$$) for the 11×11 SMU size. At extremely high cutoff grades ($z = 445$ and 854), profit figures for the smaller 11×11 SMU size are higher than those at the 22×22 SMU size, because of better selectivity (more accurate mining). At the other three cutoff grades, profit figures for the 22×22 SMU size are higher than those for the 11×11 SMU size, because of the lower cost C_0.

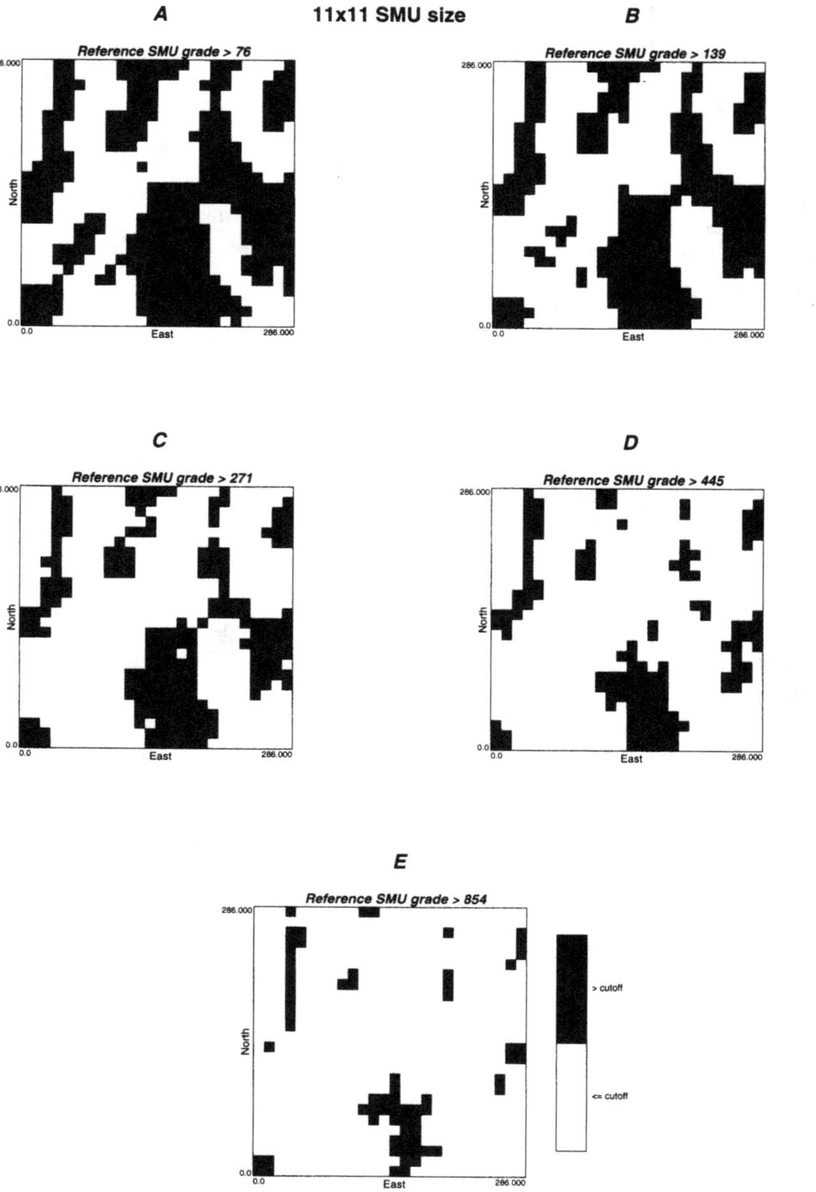

Figure IV.7: Maps of reference recovery indicators under perfect selection (SMU size 11×11).

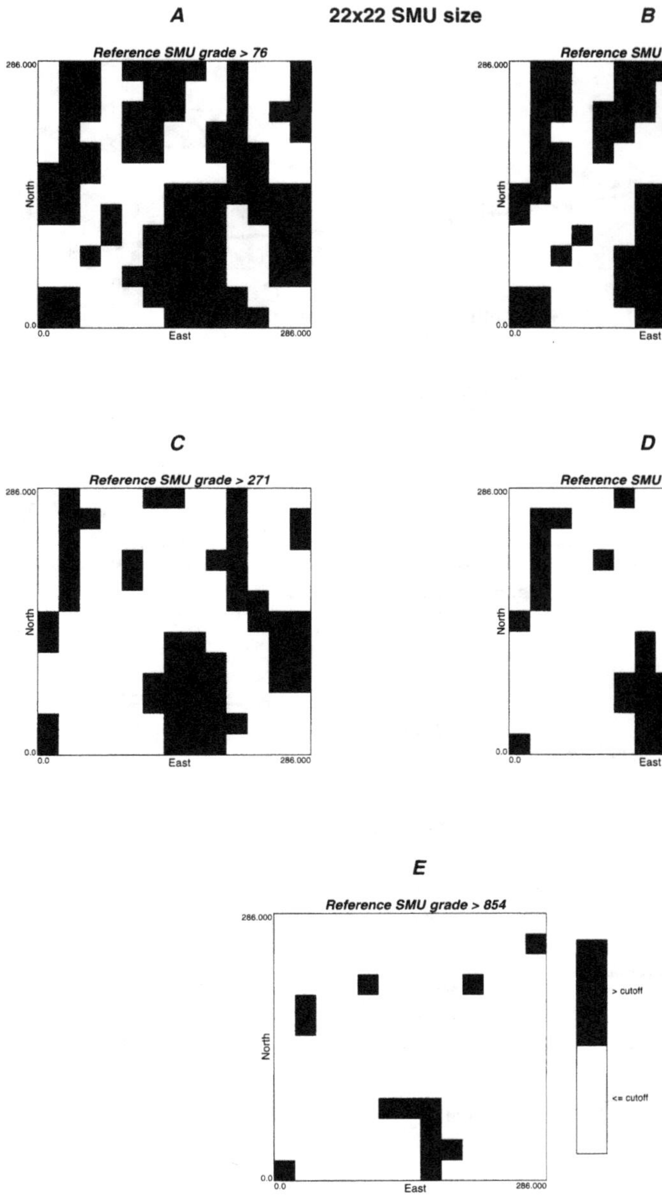

Figure IV.8: Maps of reference recovery indicators under perfect selection (SMU size 22×22).

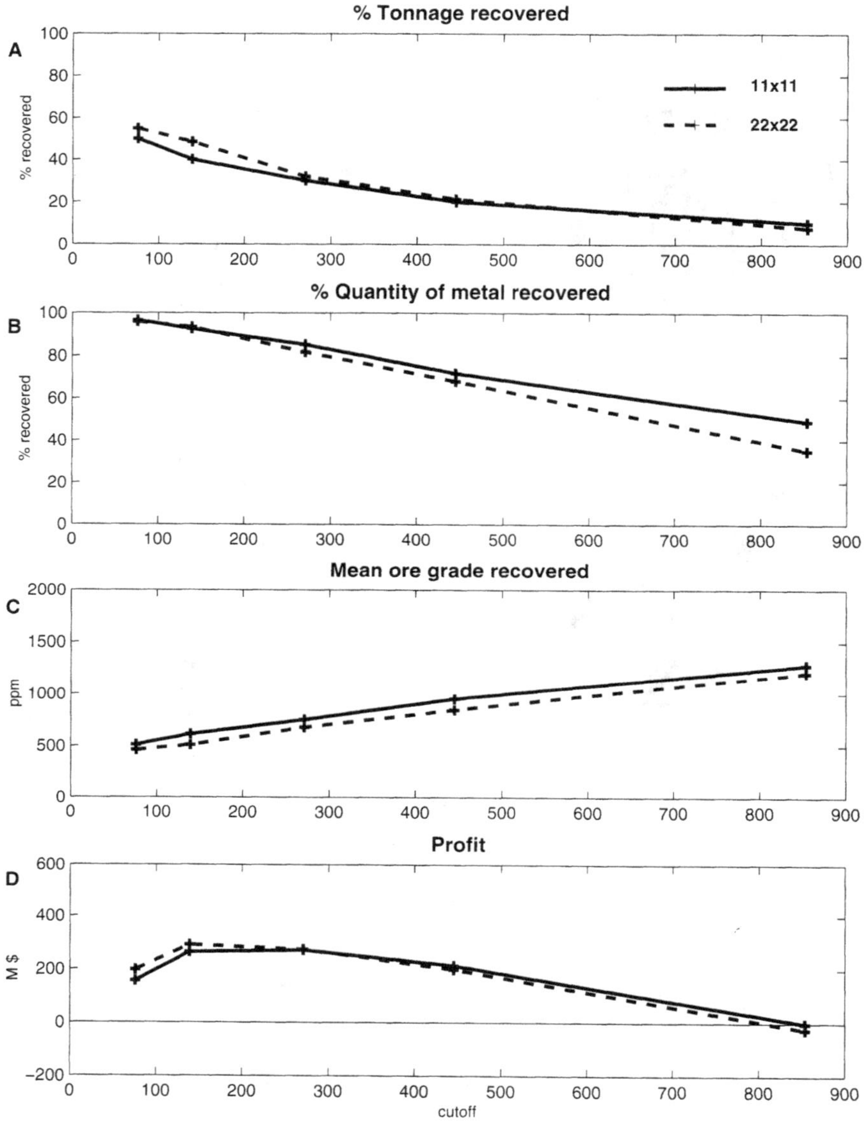

Figure IV.9: Reference recovery functions under perfect selection (Ely data set). **A:** percent tonnage recovered (total tonnage is for zero cutoff), **B:** percent quantity of metal recovered, **C:** mean ore grade, **D:** profit.

IV.3 Reference Actual Selection

As discussed in Section I.3, actual selection will be performed on future estimated SMU grades $z_v^{**}(\mathbf{u})$ based on data $b(\mathbf{u}'_\alpha)$ that will be available on a denser grid than the presently available diamond drillhole exploration data $z(\mathbf{u}_\alpha)$. Such future data are usually of poorer quality, say, blasthole data in an open pit situation or channel or long hole samples in an underground mine; they are denoted as $b(\mathbf{u}'_\alpha) \neq z(\mathbf{u}'_\alpha)$.

At the development stage of a new mine, it is usually difficult to predict the density and above all the quality of the data that will be available at the time of selection, and consequently to predict the impact on recovery and profit of the "information effect". Such prediction calls for a simulation of the future poorer quality data $b(\mathbf{u}'_\alpha)$. In this section, these future samples $b(\mathbf{u}'_\alpha)$ are simulated by adding a measurement error to the "true" reference values $z(\mathbf{u}'_\alpha)$. In a real case situation, the "true" values $z(\mathbf{u})$ are not available, and hence they will have to be replaced by simulated values $z^{(l)}(\mathbf{u})$, with the superscript (l) denoting the lth simulated realization. As a consequence, the future data would also be simulated and denoted as $b^{(l)}(\mathbf{u}'_\alpha)$ [see Section V.3].

The critical aspect of any prediction of the impact of the "information effect" is the simulation of the future data available at the actual time of selection. If statistics of such future data are available either from the (operating) mine under study or from a deemed similar operation, these statistics should be used. In the following we chose the more difficult route of simulating the future data $b(\mathbf{u}'_\alpha)$ from various considerations based on experience gained from observed differences between diamond drill hole (DDH) data assumed exact, and large diameter blasthole (BH) data of significantly poorer quality.

IV.3.1 Blasthole Data Generation

The idea is to add an error $r(\mathbf{u})$ to the reference (true) point support grade $z(\mathbf{u})$. That error depends on the local characteristics of the z-distribution around the location $\mathbf{u}$; that is, the error is heteroscedastic (see Deutsch and Journel, 1998, p. 95). The blasthole value is then written as:

$$b(\mathbf{u}) = \max\{0, z(\mathbf{u}) + r(\mathbf{u})\}, \quad \mathbf{u} \in D \tag{IV.6}$$

where $r(\mathbf{u})$ follows a normal distribution $N(m_R(\mathbf{u}), \sigma_R^2(\mathbf{u}))$, with $Cov\{R(\mathbf{u}), Z(\mathbf{u})\} \neq 0$.

The local error $r(\mathbf{u})$ is therefore a realization of a nonstationary random variable $R(\mathbf{u})$, normally distributed with local mean $m_R(\mathbf{u}) \neq 0$ and heteroscedastic local variance $\sigma_R^2(\mathbf{u})$.

The error mean $m_R(\mathbf{u})$ is modeled as a location-dependent function of the true value or signal $z(\mathbf{u})$, through a relative deviation parameter $\lambda(\mathbf{u})$:

$$m_R(\mathbf{u}) = \lambda(\mathbf{u}) z(\mathbf{u}), \quad \mathbf{u} \in D \tag{IV.7}$$

The relative deviation factor $\lambda(\mathbf{u})$ depends on the level of the signal values $z(\mathbf{u}')$ within a neighborhood $W(\mathbf{u})$ centered at location $\mathbf{u}$. That level is measured by the neighborhood mean $m_W(\mathbf{u})$ of z-values defined as:

$$m_W(\mathbf{u}) = \frac{1}{|W|}\int_{W(\mathbf{u})} z(\mathbf{u}')d\mathbf{u}' = \frac{1}{n_W}\sum_{j=1}^{n_W} z(\mathbf{u}_j), \quad \mathbf{u}_j \in W(\mathbf{u})$$

where $n_W = 25$ is the number of signal values $z(\mathbf{u}')$ within each neighborhood $W(\mathbf{u})$ of constant size $|W| = 5 \times 5$ units.

The factor $\lambda(\mathbf{u})$ is determined from the neighborhood mean value $m_W(\mathbf{u})$ by reading from Table IV.2. More precisely, $\lambda(\mathbf{u})$ is determined by the following algorithm:

1. Establish the histogram of all 286×286 mean values $\{m_W(\mathbf{u}), \mathbf{u} \in D\}$. Border effects are dealt with appropriately. We use here the true values $\{z(\mathbf{u}'), \mathbf{u}' \in D\}$, but simulated z-values could be used.

2. Determine the four quintiles, $\{F^{-1}_{m_W}(p), p = 0.2, 0.4, 0.6, 0.8\}$, of that histogram. These quantiles define five equal frequency classes of neighborhood mean values $m_W(\mathbf{u})$.

3. According to which class $m_W(\mathbf{u})$ pertains, read the correction factor from Table IV.2; for example:

    - $\lambda(\mathbf{u}) = 3.0$ for the lower class $m_W(\mathbf{u}) < 4$, leading to a large positive bias: on average $b(\mathbf{u}) > z(\mathbf{u})$ when $z(\mathbf{u})$ is located in a poor grade neighborhood.

    - $\lambda(\mathbf{u}) = -0.2$ for the higher class $m_W(\mathbf{u}) > 432$, leading to a small relative but large absolute negative bias: on average $b(\mathbf{u}) < z(\mathbf{u})$ when $z(\mathbf{u})$ is located in a high-grade neighborhood.

In short, because of averaging on a larger volume support and because of poor material recovery, blasthole b-grades tend to overestimate most z-grades and underestimate the very high z-grades.

$m_W(\mathbf{u})$	$\lambda(\mathbf{u})$	$s_W(\mathbf{u})$	$\nu(\mathbf{u})$
<4	3.0	<1	1.8
4–31	1.9	1–8	0.9
31–127	0.5	8–37	0.5
127–432	0.1	37–117	0.9
>432	−0.2	>117	2.0

Table IV.2: Parameters $\lambda(\mathbf{u})$ and $\nu(\mathbf{u})$ for the normal distribution of error at location $\mathbf{u}$ determined from the neighborhood mean $m_W(\mathbf{u})$ and standard deviation $s_W(\mathbf{u})$ values.

The error variance $\sigma_R^2(\mathbf{u})$ depends on the error mean $m_R(\mathbf{u})$ and the local variance of the z-signal through the spatially variable coefficient of variation:

$$\frac{\sigma_R(\mathbf{u})}{m_R(\mathbf{u})} = \nu(\mathbf{u}), \quad \mathbf{u} \in D \tag{IV.8}$$

where the coefficient $\nu(\mathbf{u})$ is a function of the neighborhood variance $s_W^2(\mathbf{u})$ defined as:

$$s_W^2(\mathbf{u}) = \frac{1}{n_W} \sum_{j=1}^{n_W} [z(\mathbf{u}_j) - m_W(\mathbf{u})]^2, \quad \mathbf{u}_j \in W(\mathbf{u})$$

The coefficient of variation $\nu(\mathbf{u})$ is read from Table IV.2, using the following steps:

1. Establish the histogram of all 286×286 neighborhood standard deviation values $\{s_W(\mathbf{u}), \mathbf{u} \in D\}$.

2. Determine the four quintiles of that histogram.

3. According to which quintile class $s_W(\mathbf{u})$ pertains, read the coefficient $\nu(\mathbf{u})$ from Table IV.2; for example:

   - $\nu(\mathbf{u}) = 1.8$ for the lower class $s_W(\mathbf{u}) < 1$, leading to a large error coefficient of variation,

   - $\nu(\mathbf{u}) = 0.5$ for the median class $s_W(\mathbf{u}) \in (8.0, 37.0]$, leading to a much lower error coefficient of variation.

In short, the error variance is greater in areas of very low and very high neighborhood variance values. Also, the greater the mean error $m_R(\mathbf{u})$, the greater the error variance $\sigma_R^2(\mathbf{u})$.

The error structure generated by expressions (IV.7) and (IV.8) is far different from the usually too congenial error models, which assume independence of the error from the signal: the local statistics $m_R(\mathbf{u})$ and $\sigma_R^2(\mathbf{u})$ are then assumed stationary. Although purely synthetic at present, we believe our error simulation procedure to be more realistic.

In any particular application the values of Table IV.2 or expressions (IV.7 and IV.8) could be made different based, hopefully, on some calibration data. For example, all λ-values can be made positive, corresponding to overestimation at all range of grades.

Results

Figure IV.10 gives the grayscale plots of the neighborhood mean $m_W(\mathbf{u})$ and standard deviation $s_W(\mathbf{u})$ values, along with their corresponding histograms and scattergram. Note the proportional effect (Journel and Huijbregts, 1978, p. 186): $s_W(\mathbf{u})$ increases linearly with $m_W(\mathbf{u})$ [see Fig. IV.10E]. The grayscale plot of the error field and its histogram are given in Figures IV.11A and

IV.11B. Note that although each local error distribution is Gaussian per the error model, the resulting global histogram of error is not anymore Gaussian, or even symmetric; in addition, if the mean and median error are small, some very large errors may occur. Figure IV.11C gives the scattergram of the error $r(\mathbf{u})$ versus the signal $z(\mathbf{u})$ displaying clearly the dependence of the error on the signal. The negative correlation (-0.40) indicates that large z-values tend to be underestimated and that low z-values are overestimated (compare also Fig. IV.11A to the z-reference map of Fig. IV.1A).

The exhaustive error r-field of Figure IV.11A is added to the exhaustive reference z-field to generate the poorer-quality (blasthole) data b-field of Figure IV.12A, following expression (IV.6). Figure IV.12A provides a blurred and noisy image of the reference field of Figure IV.1A, yet the correlation between collocated $b(\mathbf{u})$ and $z(\mathbf{u})$ values remains high at 0.89 (see scattergram of Fig. IV.12C). Note that the correlation is poorer at both extremes of the z-distribution. The histogram of b-values (Fig. IV.12B), when compared to the reference z-histogram (Fig. IV.1B), shows a slight bias and lesser spread, better seen from the reduced interquartile range $[18, 302]$ versus the reference $[7, 317]$. Notwithstanding that overall lesser spread, because of a few very large errors, some extremely large b-values exceed the corresponding z-values: the maximum b-value is 11853 versus the z-maximum of 6740. Underestimation of high z-signal values is typical of many measurement devices (e.g., blasthole gold grades), whereas a few extremely large measurements may be the result of various device failures (again not uncommon in gold grade sampling). Figure IV.12D gives the standardized semivariograms of the blasthole data along the two principal directions N10°E and N80°W, with the fit by the following anisotropic model:

$$\gamma_B(\mathbf{h}) = 0.15 + 0.45\text{Exp}\sqrt{\left(\frac{h_1}{50}\right)^2 + \left(\frac{h_2}{30}\right)^2} + 0.4\text{Exp}\sqrt{\left(\frac{h_1}{100}\right)^2 + \left(\frac{h_2}{40}\right)^2} \tag{IV.9}$$

When compared to the reference Z-semivariogram model $\gamma_Z(\mathbf{h})$ given in expression (IV.2), this blasthole semivariogram model $\gamma_B(\mathbf{h})$ reflects a higher nugget effect and shorter ranges for the first exponential structure. Again, this would be expected from poorer-quality data (blasthole data).

Future blasthole grid

Now that the exhaustive blasthole b-field (Fig. IV.12A) has been generated, we can sample it retaining only the grid of b-sample grades that are expected to be available at the time of selection.

Figure IV.13 gives the maps and statistics of the blasthole grades on the 22×22 grid retained. The same grid was used for estimating the SMU grades for both 11×11 and 22×22 sizes. Note that the sample grid (Fig. IV.13A) suffices to detect the main reference high-grade structures of Figure IV.1A. Similarly and notwithstanding the poorer quality of the blasthole data, the histogram of Figure IV.13B reflects reasonably well the reference histogram of

Figure IV.1B. It is remarkable how data quantity can remediate data quality, provided the latter is not too poor.

Figure IV.13C gives the standardized semivariograms of the sample blasthole data, along with the model fit. When compared to the exhaustive semivariogram model $\gamma_B(\mathbf{h})$ given in expression (IV.9), the semivariograms calculated from the coarse 22×22 grid exhibit a larger nugget effect and larger ranges; the following semivariogram model $\gamma_B^f(\mathbf{h})$ was adopted:

$$\gamma_B^f(\mathbf{h}) = 0.25 + 0.35\text{Exp}\sqrt{\left(\frac{h_1}{100}\right)^2 + \left(\frac{h_2}{40}\right)^2} + 0.40\text{Exp}\sqrt{\left(\frac{h_1}{200}\right)^2 + \left(\frac{h_2}{50}\right)^2} \tag{IV.10}$$

where the superscript f denotes that this model was inferred from future blasthole grades on the 22×22 grid.

Comparing model (IV.10) fitted from a sample of blasthole values to the exhaustive model (IV.2), one realizes how much noise sparse and poor-quality data can add to the underlying "true" variogram.

IV.3.2 Kriging from Future Blasthole Data

The information available at the time of selection consists of the blasthole grade b-data on the 22×22 grid. Each SMU grade is estimated from these b-data, and selection is performed on the resulting estimated SMU grades $z_v^{**}(\mathbf{u})$ [see relations (I.6) to (I.10)].

Ordinary kriging was adopted for estimating each SMU grade using the nearest nine blasthole data $b(\mathbf{u'}_\alpha)$; that is, the central b-datum and the eight b-data of the first aureola in the case of a 22×22 SMU. Fewer data were used for SMUs located at the borders of the deposit. The semivariogram model $\gamma_B^f(\mathbf{h})$, given in relation (IV.10), was considered for block kriging performed with a 4×4 discretization of each 11×11 block (SMU); the GSLIB program `kt3d` was used for this purpose.

Figure IV.14 gives the grayscale plots of the resulting kriging estimates and their statistics. As expected, the kriging maps and histograms display the characteristic smoothing effect (compare to the reference grade maps of Figs. IV.5C–F). Remarkably, the main reference high-grade structures are accurately located, although with variance smoothing. Figures IV.14E and IV.14F give the scattergrams of the reference true SMU grades versus their kriging estimates: The correlation is high at 0.75 and 0.85, yet some conditional bias is apparent—overestimation of low reference values and underestimation of high reference values.

In a real case situation, true SMU grades are rarely, if ever, available to allow a comparison of the type made in Figures IV.14E and IV.14F. Also (at least in precious metal mining operations), one would not expect such high estimation accuracy scores (a 0.85 correlation). This leads us to believe that notwithstanding our effort to generate realistic error distributions (see Table IV.2), the resulting blasthole grades $b(\mathbf{u})$ are still too "good" in comparison to what is available in practice. Alternatively, our reference z-field (Fig. IV.1A) based on elevation data is too congenial, too well-behaved, to mimic adequately the complexity of even a pervasive low-grade precious metal deposit. Indeed, the reference statistics of Figure IV.1B and Figure IV.2 indicate a single mode distribution and a zero nugget effect, extremely well-behaved (stationary), semivariogram.

IV.3.3 Actual Selection Based on Kriging Estimates

Selection is performed based on the SMU grades $z_v^{**}(\mathbf{u})$ estimated by block kriging (see Figs. IV.14A and IV.14B). The resulting selection indicator maps are given in Figure IV.15 for the 11×11 SMU size and in Figure IV.16 for the 22×22 SMU size.

These actual recovery maps should be compared to the ideal recovery maps of Figure IV.7 and Figure IV.8, corresponding to perfect selection performed on the true SMU grades $z_v(\mathbf{u})$. For the 11×11 SMU, the overall match is excellent except at the highest 90th-percentile cutoff value $z = 854$. The coarser 22×22 SMU size entails a significant loss of resolution already visible at the second cutoff value $z = 139$ (compare Fig. IV.16 to its reference counterpart Fig. IV.8).

Figures IV.15 and IV.16 provide the spatial distribution of the actual tonnage recovered under imperfect selection. The corresponding SMU true block grades $z_v(\mathbf{u})$ are recovered, defining the actual recovered quantities of metal and mean ore grades [see expressions (I.8) and (I.9)]. Table IV.3 gives these actual recovery figures (tons, quantity of metal, mean ore grade, and profit) for the five cutoff values retained. Figure IV.17 gives a graphical representation of these recovery values as functions of the cutoff z.

As expected, the tonnage and quantity of metal recovered decrease as the threshold increases (Figs. IV.17A and IV.17B), whereas the reverse is true for the mean ore grade recovered (Fig. IV.17C). Differences between tonnage recovered at the 11×11 and 22×22 SMU sizes are very small—hardly discernible at the plot resolution. Maximum profit is reached at the second cutoff grade for both 11×11 and 22×22 SMU sizes: $222M\$$ and $262M\$$, respectively (see Fig. IV.17D). The profit figures at the 22×22 SMU size are always higher than those at the 11×11 SMU size, because of better selectivity (more accurate mining) being offset by higher operational costs for the latter. Note that the profit is negative for both SMU sizes at the highest cutoff grade $z = 854$.

The difference between the actual recovery characteristics (see Fig. IV.17 and Table IV.3) and the perfect selection recovery (see Fig. IV.9 and Table IV.1) quantifies the impact of imperfect selection; that is, the impact of the "information effect". To facilitate comparison, the results of Tables IV.1 and IV.3 have been concatenated in Figure IV.18, which gives the relative difference between actual and perfect recovery as a function of the cutoff z.

11 × 11 SMU				
Cutoff (z)	% Tonnage	% Q. of Metal	Mean Grade	Profit
76	69.2	98.3	376	140
139	52.5	92.5	467	222
271	28.6	73.0	677	179
445	15.4	54.4	937	110
854	4.0	18.2	1205	-137
22 × 22 SMU				
Cutoff (z)	% Tonnage	% Q. of Metal	Mean Grade	Profit
76	74.0	98.0	351	183
139	49.7	89.4	477	262
271	27.8	71.1	678	220
445	16.6	54.5	872	139
854	3.6	16.4	1226	-91

Table IV.3: Reference recovery values (Ely data set) for SMU sizes 11 × 11 and 22 × 22 under actual selection. % tonnage refers to total tons = 286 × 286 = 81796 unit tons (at zero cutoff), % quantity of metal refers to total = 81796 × 265 unit tons × grade unit (265 ppm is the overall mean grade), and profit is given in $M\$$.

For tonnage recovery, for example, that relative difference is written as:

$$\Delta t_v(z) = \frac{t_v^*(z) - t_v^0(z)}{t_v^0(z)}$$

where $t_v^0(z)$ is the ideal recovery under perfect selection at cutoff z, and $t_v^*(z)$ is the actual recovery with selection performed on the estimated SMU grade $z_v^{**}(\mathbf{u})$.

Similarly, relative difference values can be defined for the quantity of metal recovered $\Delta q_v(z)$ and the mean ore grade recovered $\Delta m_v(z)$. Note that any of these relative differences may be positive. The relative difference in profit, however, is always negative and is plotted as an absolute relative profit loss $|\Delta p_v(z)|$. Indeed, the actual profit cannot exceed the ideal profit under perfect selection, that is:

$$p_v^*(z) \leq p_v^0(z), \quad \forall\ z,\ \forall\ \text{SMU size } |v| \tag{IV.11}$$

From Figure IV.18, it appears that beyond the second cutoff ($z = 139$), actual tonnage and quantity of metal are significantly lesser than what perfect selection would provide, with the mean ore grade being approximately the same. This lesser quantity of ore leads to a significant profit loss; for example, 34% or $93M\$$ (in dollar terms) for the third cutoff $z = 271$ and SMU size 11×11. The relative profit loss at the extreme cutoff grade ($z = 854$), not shown at the graph scale, is as high as 6750%. Indeed, improving the accuracy of ore selection is of paramount importance in highly selective mining operations.

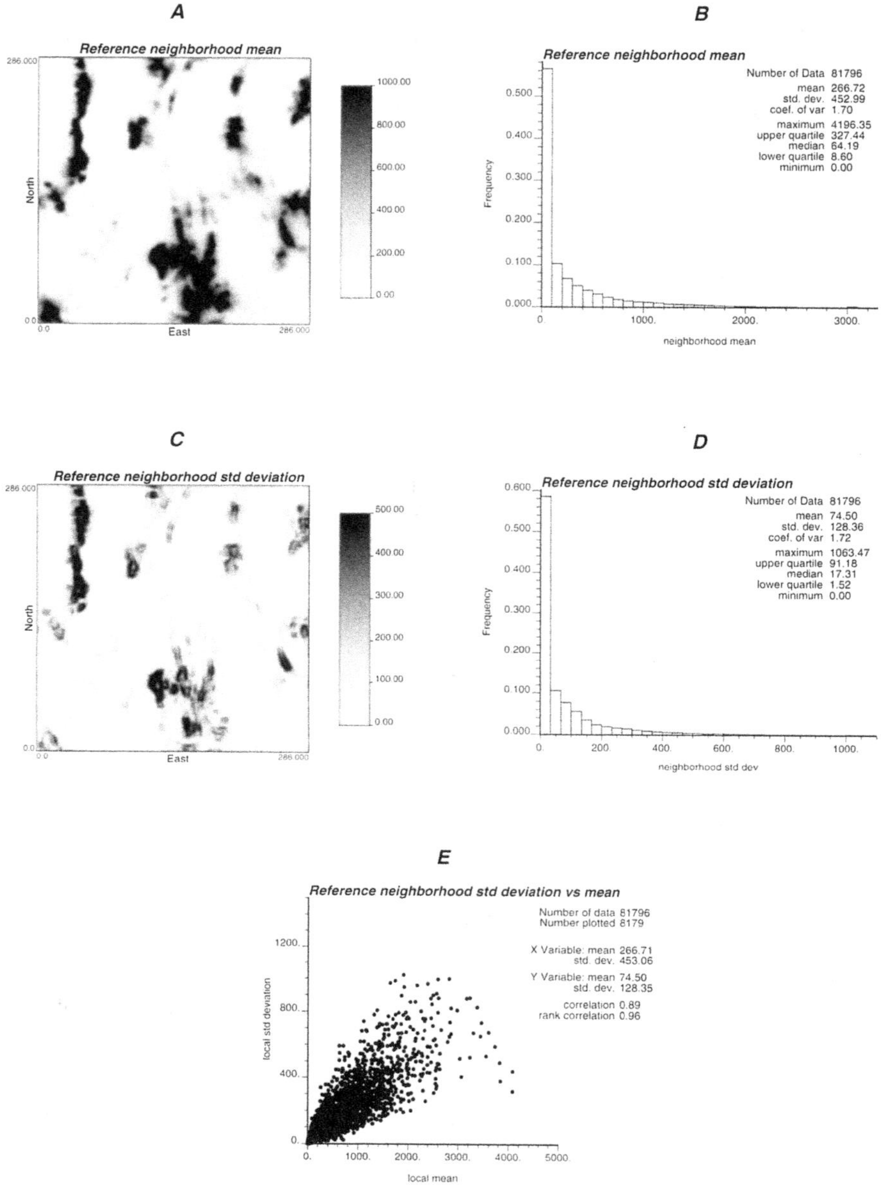

Figure IV.10: Reference local neighborhood statistics used to define the local error statistics (neighborhood size is 5 × 5 units). **A**, **B:** local mean of reference values and corresponding histogram, **C**, **D:** local standard deviation of reference values and corresponding histogram, **E:** scattergram of local mean versus local standard deviation values (only a few points are plotted on this scattergram to avoid cluttering, but the statistics are calculated from all 81796 data pairs).

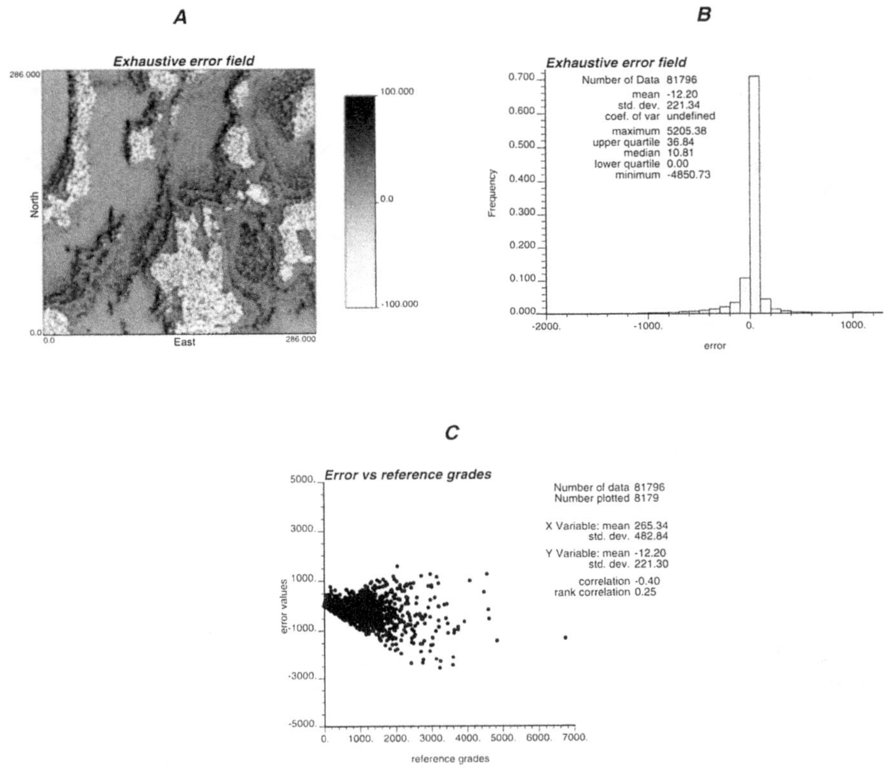

Figure IV.11: Statistics of the error r-field to be added to the reference z-grades for generating the poorer-quality (future blasthole) b-data. **A:** pixel plot of the error r-values $[r(\mathbf{u}) = b(\mathbf{u}) - z(\mathbf{u})]$, **B:** histogram of r-values, **C:** scattergram of r- versus true z-values (note the error heteroscedasticity). Although only a few points are plotted, the statistics are calculated from all 81796 pairs.

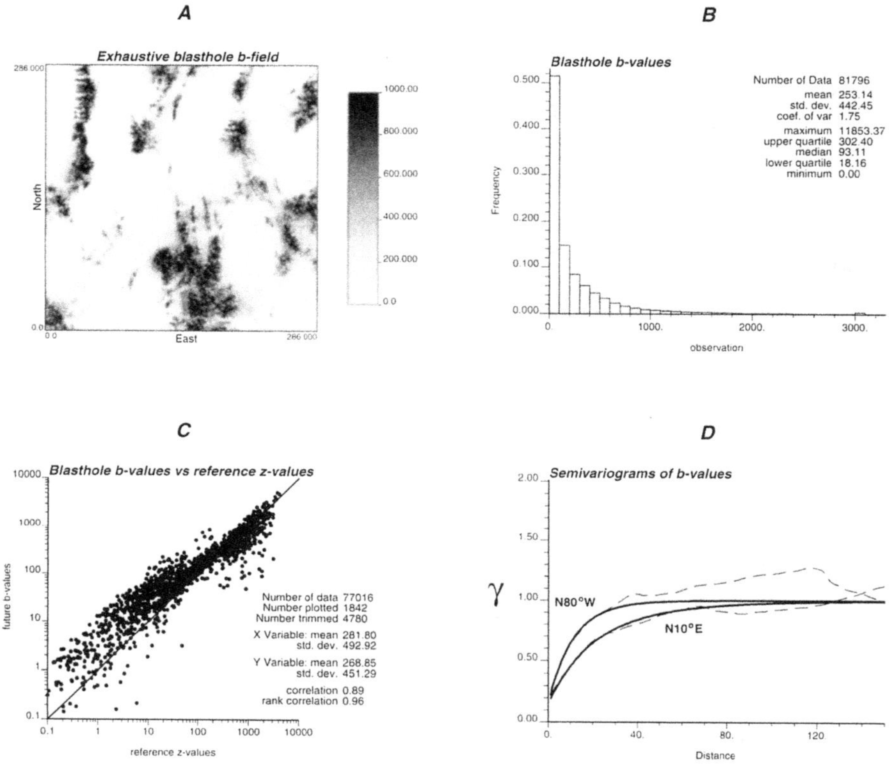

Figure IV.12: Statistics of the exhaustive poorer-quality blasthole data field. **A:** grayscale plot of blasthole b-data, **B:** histogram of b-values, **C:** scattergram of b-values versus reference z-values (log-scale) where zero values have been trimmed out, **D:** standardized b-semivariograms along N10°E, N80°W, and model fit.

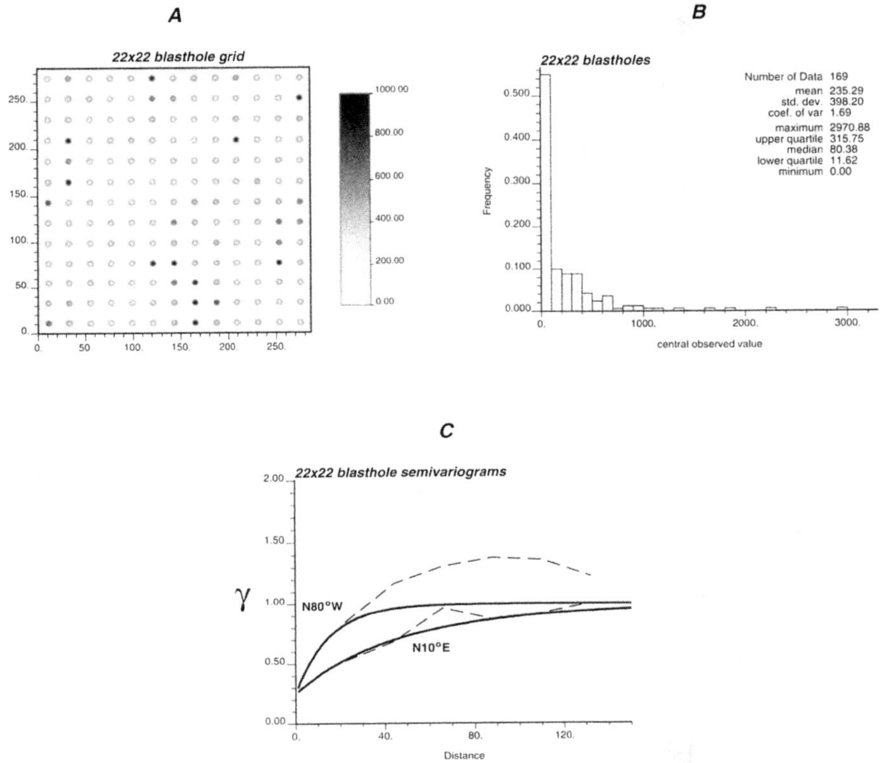

Figure IV.13: Future sample blasthole data on a 22 × 22 grid and corresponding statistics. **A:** location map of future b-samples, **B:** histogram of b-samples, **C:** standardized semivariograms of b-values along N10°E, N80°W, and model fit.

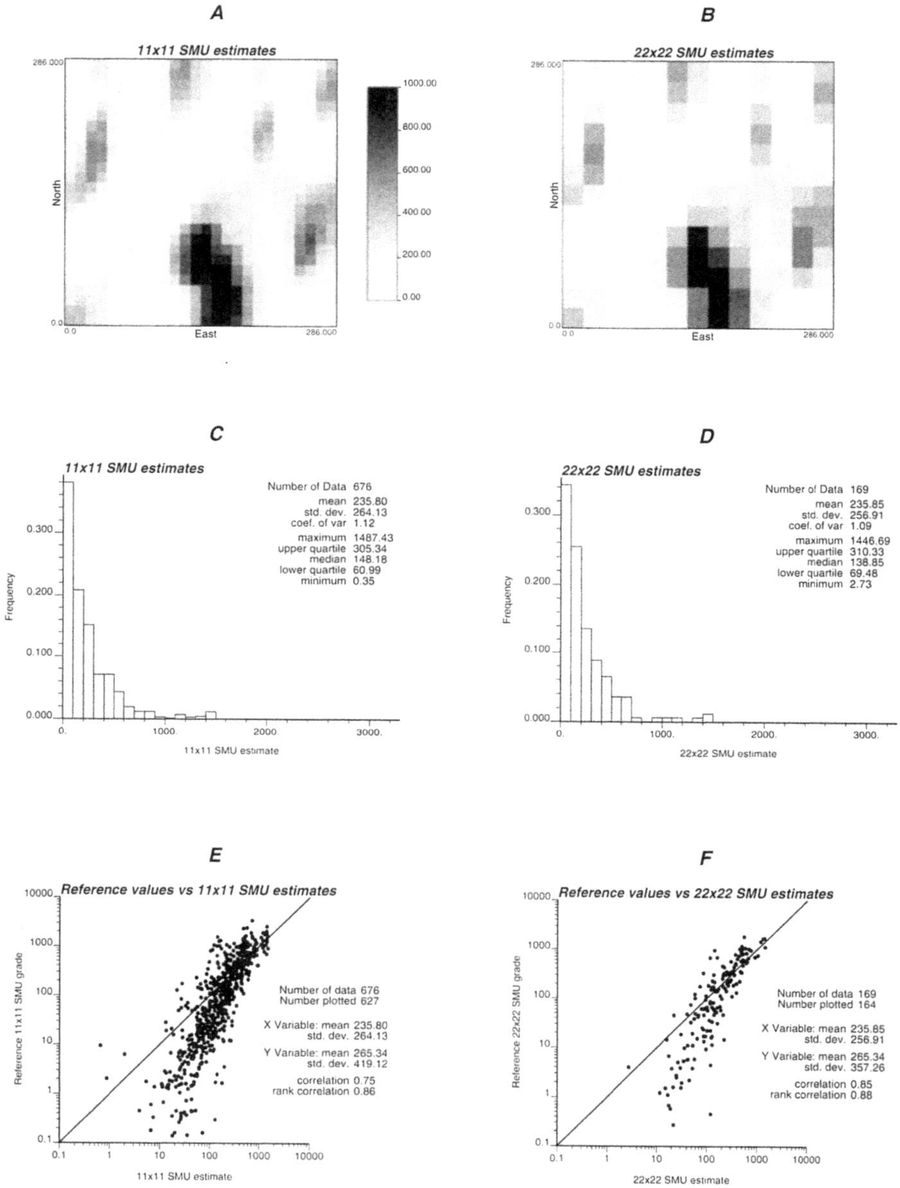

Figure IV.14: Statistics of block support grade estimates of SMU size 11×11 (left) and 22×22 (right), obtained via ordinary kriging using future blasthole data. **A**, **B:** grayscale plots, **C**, **D:** histograms, **E**, **F:** scattergrams of true versus estimated grades. Zero values are not plotted in the scattergrams, but were accounted for in the statistics.

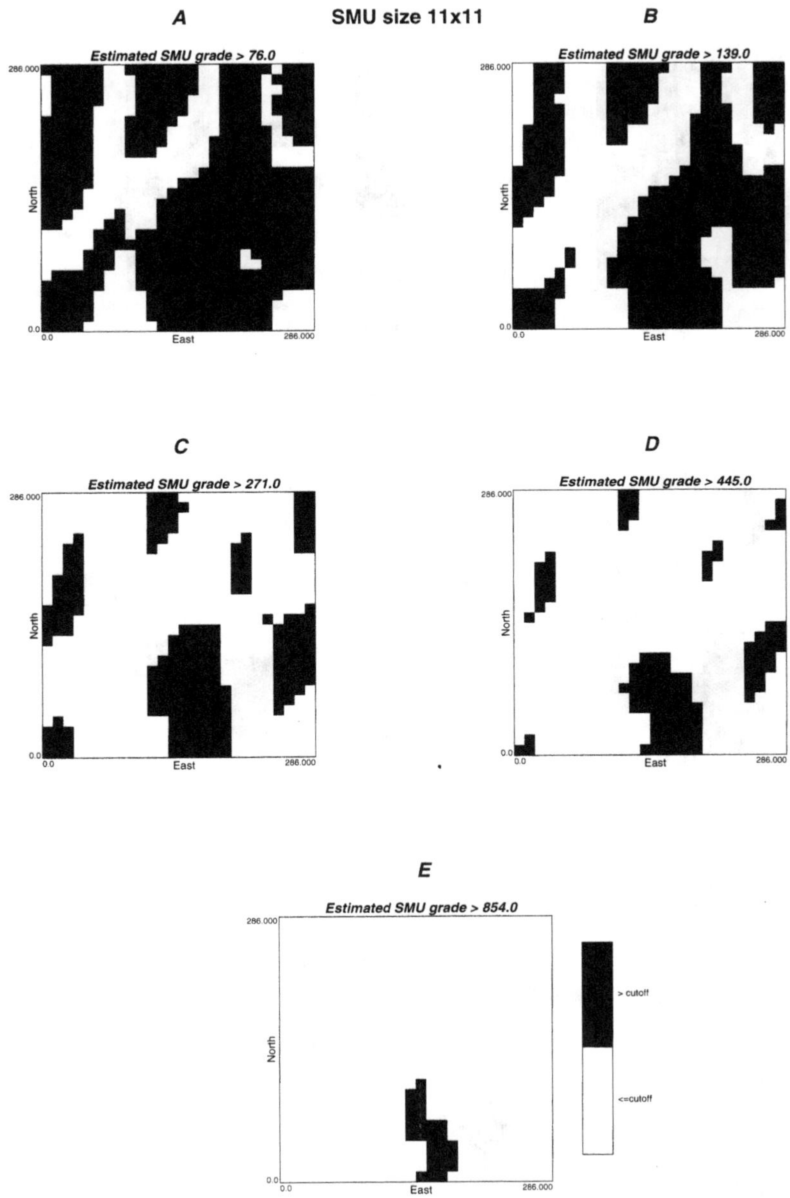

Figure IV.15: Maps of reference recovery indicators under actual selection performed on estimated SMU grades (SMU size 11×11).

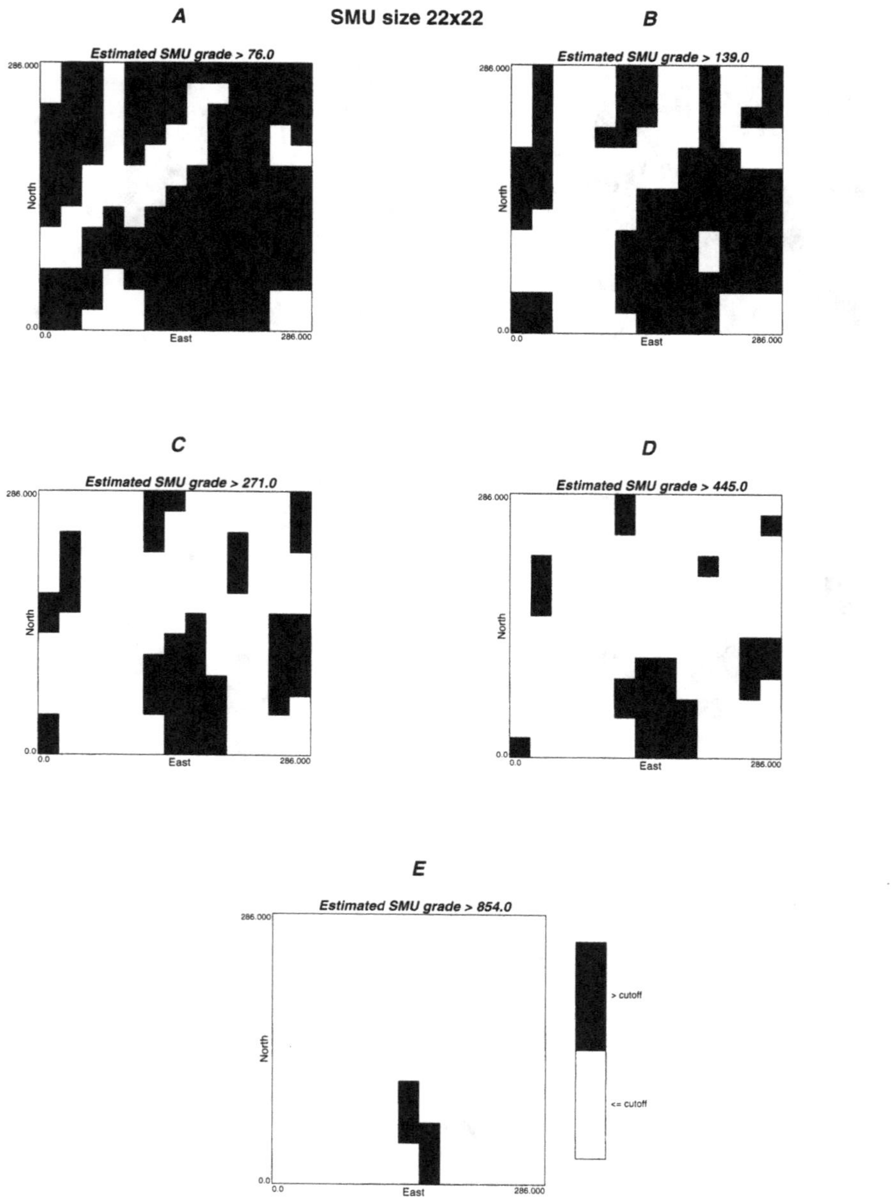

Figure IV.16: Maps of reference recovery indicators under actual selection performed on estimated SMU grades (SMU size 22×22).

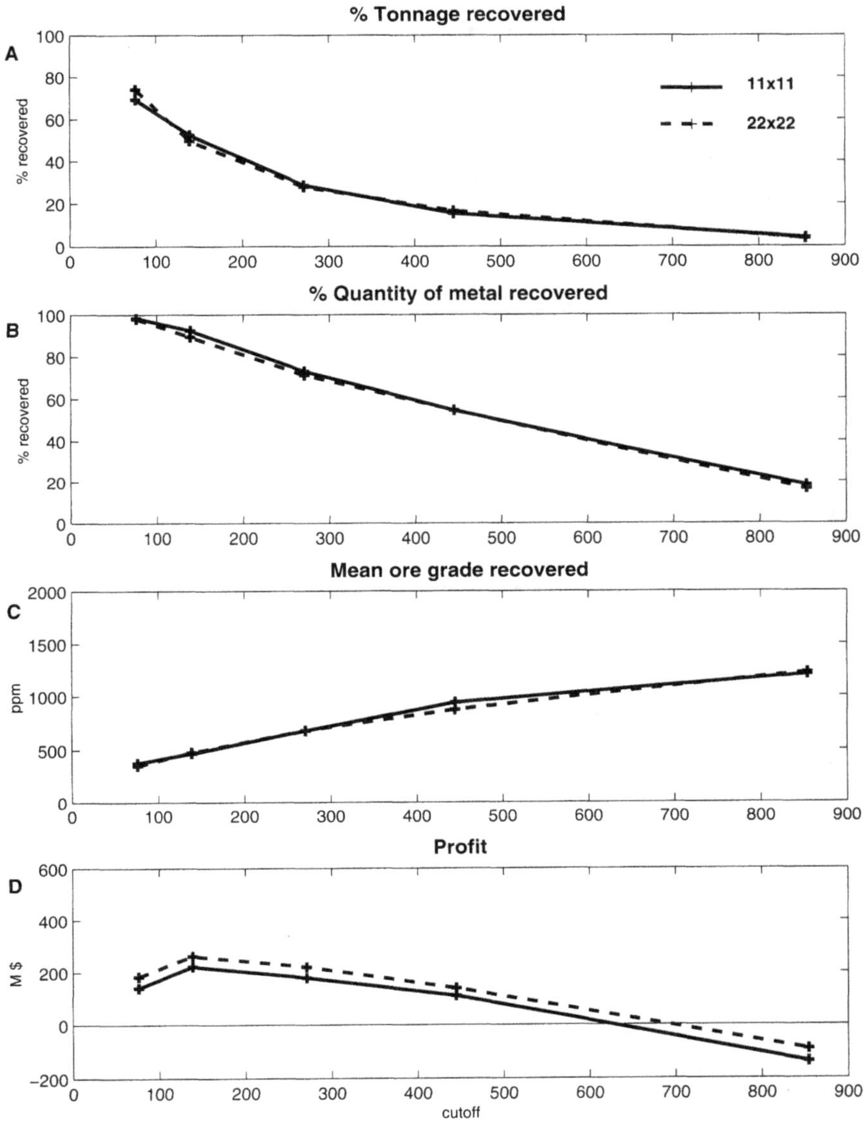

Figure IV.17: Reference recovery functions under actual selection (Ely data set) based on estimated SMU grades using imperfect future data (the corresponding true SMU grades are recovered). **A:** percent tonnage recovered (total tonnage is for zero cutoff), **B:** percent quantity of metal recovered, **C:** mean ore grade, **D:** profit.

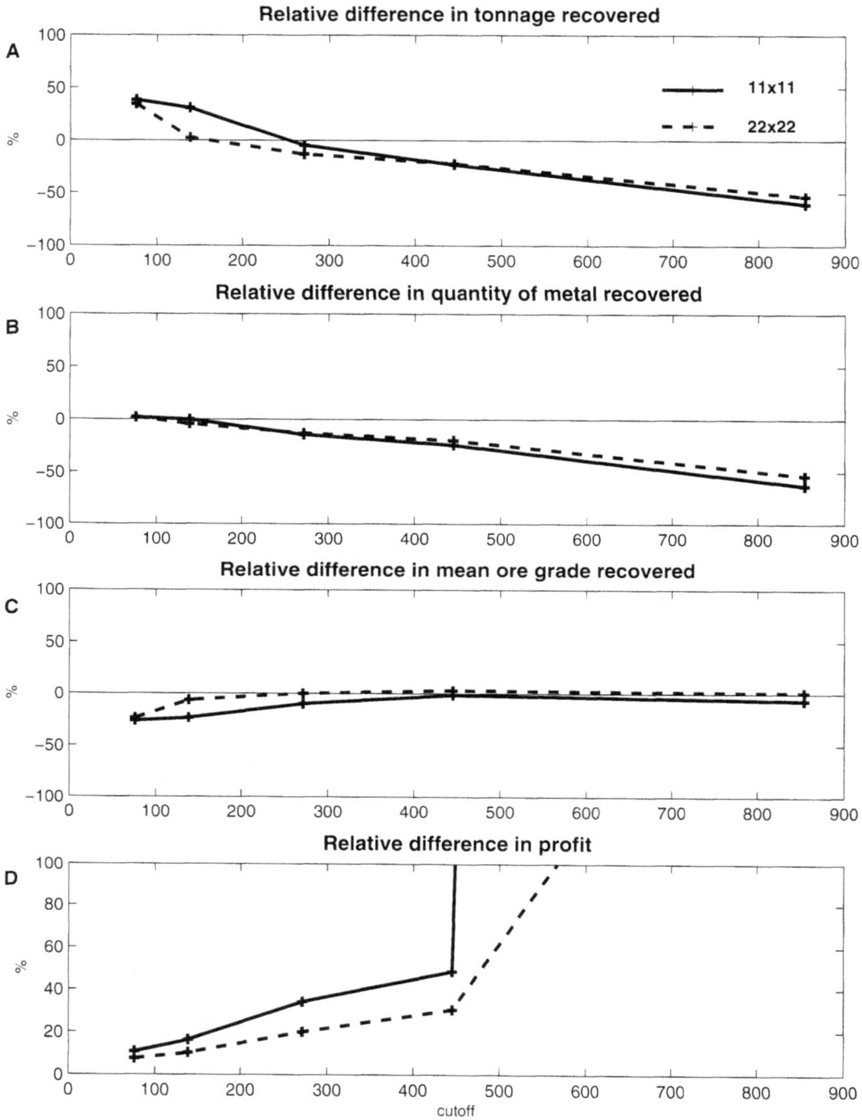

Figure IV.18: Relative difference in reference recovery functions under actual and perfect selection (profit under actual selection is always smaller than under perfect selection, and hence the absolute relative difference is given).

IV.4 The Ely2 Reference Data Set

The original Ely elevation data set proved to be too congenial in its spatial variability, displaying a single reasonably multiGaussian population after normal score transform (recall Figs. IV.1–IV.4). Were the coefficient of variation of Figure IV.1B reduced to 0.3, the spatial variability of this data set would be more representative of BTU grades in a coal seam than gold grades in a Carlin-type deposit (Nevada). Hence the idea of further transforming the Ely data set to generate a two-population distribution with a marked difference between the spatial structures of a pervasive low-to-median-grade population and a very high-grade population with clear anisotropy that could represent vein-related gold enrichment. Again, an actual data set would have been preferable. Because such an exhaustive gridded reference data set does not exist, we had to create it starting from a "natural" distribution, in this case the Ely topographic elevation data.

IV.4.1 Generation of the Ely2 Reference Data

Starting from the previous Ely reference data set displayed in Figure IV.1, the following algorithm was implemented to generate a second high-grade population with marked anisotropy:

1. Apply the threshold $z_0 = 1000$, thus creating two sets of values: $z^+(\mathbf{u}) = z(\mathbf{u}) > z_0$, and $z^-(\mathbf{u}) = z(\mathbf{u}) \leq z_0$. The threshold z_0 corresponds to the 93rd percentile of the marginal distribution of the Ely data set; thus, the proportion of values $z^+(\mathbf{u})$ is 0.07.

2. To create a marked, yet gradual, difference between grades in the two populations, the following transform was applied to enhance the z^+ data only:

$$\text{new } z^+(\mathbf{u}) = z^+(\mathbf{u}) \cdot \exp\left[(z^+(\mathbf{u}) - z_0)/2000\right]$$

   z^+-values close to the threshold z_0 are not affected significantly by the transform, whereas high original z^+ values are considerably amplified. For example: if $z^+(\mathbf{u}) = 1500$, then new $z^+(\mathbf{u}) = 1500\exp(0.25) = 1926$. If $z^+(\mathbf{u}) = z_{max} = 6740$, then new $z_{max} = 6740\exp(2.87) = 118874$.

3. To enhance the N10°E anisotropy of the z^+ data only, an elliptical moving window elongated 15 grid units along N10°E and 3 grid units across was centered on all z^+ valued nodes, and the central value $z^+(\mathbf{u})$ was replaced by the window average of z^+ values.

This algorithm resulted in the Ely2 population, still denoted as $z(\mathbf{u})$, displayed in Figure IV.19. This data set features bimodality with distinct vein-type streaks of very high grades; these streaks cannot be seen on Figure IV.19A. A selective mining operation would strive to separate that high-grade population ($> z_0 = 1000$).

Figures IV.20A and IV.20B give the Ely2 reference normal scores semivariograms calculated across both populations: note the similar nugget effect and hole effect (dip) when compared to the original Ely variograms of Figure IV.2. Figures IV.20C and IV.20D give the semivariograms of the Ely2 normal scores grades calculated within each population separately. The normal score transform, however, is performed across the two populations. The low-grade semivariograms exhibit a greater continuity than the high-grade semivariograms and fluctuate less; this lesser fluctuation is the result of the greater number of samples in the low-grade population.

Indicator variography

Figure IV.21 gives the series of directional N10°E and N80°W reference indicator semivariograms (standardized to unit sill) for the following p-quantiles:

A: $z = 7$ corresponding to $p = 0.25$ (lower quartile)

B: $z = 60$ corresponding to $p = 0.50$ (median)

C: $z = 317$ corresponding to $p = 0.75$ (upper quartile)

D: $z = 1000$ corresponding to the threshold z_0 (93rd percentile) separating the two populations

Note the higher nugget effect of the indicator variograms at the higher threshold z_0, as compared to the low-grade indicator semivariograms.

Reference SMU data

The previous reference "point" support z-data are averaged into block support SMU grades. Again the two SMU sizes are 11×11 and 22×22 (see Figs. IV.22C–F). The grayscale maps of the SMU grades are similar to those of the original Ely data set (recall Figs. IV.5C–E).

IV.4.2 Ely2: Perfect Selection

Perfect selection is performed on the reference "true" SMU grades $z_v(\mathbf{u})$ [see relations (I.1) through (I.5)].

The same five grade cutoff grades used for the original Ely data set, given in equation (IV.3), were retained. They correspond to the following p-quantiles of the Ely2 marginal distribution $F_v(z)$ of SMU grades of size $|v| = 11 \times 11$:

$$\begin{aligned} z = q_v(p) &= 76 & p &= 0.5 \\ z &= 139 & p &= 0.6 \\ z &= 271 & p &= 0.7 \\ z &= 445 & p &= 0.8 \\ z &= 854 & p &= 0.88 \end{aligned} \tag{IV.12}$$

Note that the last cutoff $z = 854$ is close to the reference threshold $z_0 = 1000$ defining the Ely2 high-grade population. The p-values between the two lists (IV.3) and (IV.12) are the same except for the last one, because the Ely2 distribution differs from the original Ely distribution only in the upper tail.

The maps of the perfect selection indicators $i_v^0(\mathbf{u}; z)$ for SMU sizes 11×11 and 22×22 are given in Figures IV.23 and IV.24. Again these indicator maps differ little from the corresponding Ely indicator maps given in Figures IV.7 and IV.8. The most discernible difference is in the case of the highest economic cutoff grade $z = 854$ in the 22×22 SMU size (compare Fig. IV.24E and IV.8E).

Table IV.4 gives the perfect recovery figures corresponding to the five previous cutoff grades given in (IV.12); the same cutoff values were applied for both SMU sizes 11×11 and 22×22. Figure IV.25 gives the corresponding recovery curves as function of the cutoff grade z. The profit formulae (IV.4 and IV.5) used for the original Ely data set were kept.

11 × 11 SMU				
Cutoff (z)	% Tonnage	% Q. of Metal	Mean Grade	Profit
76	50.0	97.4	715	323
139	40.1	94.6	866	515
271	30.0	89.2	1090	605
445	20.4	80.0	1438	629
854	11.5	65.3	2078	435
22 × 22 SMU				
Cutoff (z)	% Tonnage	% Q. of Metal	Mean Grade	Profit
76	55.0	97.0	647	364
139	47.9	94.9	727	541
271	32.5	86.7	978	603
445	22.5	77.2	1260	600
854	13.0	61.1	1722	358

Table IV.4: Ely2 recovery values under perfect selection. % tonnage refers to total tons = 286 × 286 = 81796 unit tons (at zero cutoff), % quantity of metal refers to total = 81796 × 367 unit tons × grade unit (367 is the overall mean grade), and profit is given in $M\$$.

As expected, the tonnage and quantity of metal recovered decrease as the cutoff grade increases (Figs. IV.25A and IV.25B), whereas the reverse is true for the mean ore grade recovered (Fig. IV.25C).

The tonnage recovery for Ely2 (Table IV.4) is very similar to the tonnage recovery of the original Ely deposit (Table IV.1). The other recovery figures, quantity of metal, mean ore grade, and profit, are significantly different because of the extreme high grades of the Ely2 deposit.

IV.4.3 Ely2: Actual Selection

Similar to what was done for the original Ely reference data set (recall Section IV.3), actual selection is performed on the future estimated SMU grades $z^{**}(\mathbf{u})$ based on blasthole data $b(\mathbf{u}'_\alpha)$ that will be available at the time of actual selection and mining.

Blasthole data generation

The same algorithm for blasthole data generation [recall formulae (IV.6–IV.8)] was retained. However, the λ-parameters that control the conditional bias (quality) of the blasthole b-data were chosen different to generate b-data that underestimate more severely the very high grades (see Table IV.5 and compare to Table IV.2).

For data $z(\mathbf{u})$ in a neighborhood of very high grades ($m_W(\mathbf{u}) > 432$), the relative bias factor is now $\lambda(\mathbf{u}) = -0.5$, leading to an average 50% underestimation, resulting, for example, from loss of high-grade fine particulates.

Figure IV.26 gives the grayscale plot of the resulting error field $r(\mathbf{u}) = b(\mathbf{u}) - z(\mathbf{u})$, its histogram, and its scattergram versus the true value $z(\mathbf{u})$. Note the negative error mean (-88) corresponding to an overall underestimation of the high z-grade values. Again, the error scattergram displays heteroscedasticity with a significant negative correlation (-0.65) with the true z-signal. Comparison with Figure IV.11 indicates the poorer quality of the Ely2 blasthole field.

$m_W(\mathbf{u})$	$\lambda(\mathbf{u})$	$s_W(\mathbf{u})$	$\nu(\mathbf{u})$
< 4	3.0	< 1	1.8
4 - 29	1.9	1 - 8	0.9
29-125	0.5	8 - 38	0.5
125 - 432	-0.1	38 - 119	0.9
> 432	-0.5	> 119	2.0

Table IV.5: Ely2 parameters $\lambda(\mathbf{u})$ and $\nu(\mathbf{u})$ for the generation of the reference blasthole field.

The error field of Figure IV.26A is added to the Ely2 reference z-field of Figure IV.19A to generate the exhaustive blasthole b-field of Figure IV.27A. The b-field provides a blurred image that misses the crisp continuity of the high-grade streaks seen in the reference map of Figure IV.19A. Because of the severe underestimation of high z-grades, the blasthole data histogram also misses the second mode of high z-values seen on the reference histogram of Figure IV.19D. Yet the correlation between collocated $b(\mathbf{u})$ and $z(\mathbf{u})$ values remains high at 0.80 (see scattergram of Fig. IV.27C). Figure IV.27D gives the semivariograms of the normal score transformed b-data along the two principal directions N10°E and N80°W. Note the higher nugget effect when compared to the Ely2 reference z-semivariogram of Figure IV.20B.

Future kriging from blasthole data

From the exhaustive blasthole b-field, a sample of 169 blasthole data was retained on a 22×22 grid (see Figs. IV.28A and IV.28B). This sample represents the data available at the time of actual selection and mining.

In presence of a marked two-population mineralization, mining geostatisticians would likely consider some form of indicator kriging to separate the two populations (Journel, 1983; Davis, 1984; Journel and Isaaks, 1984). Given the sample histogram of Figure IV.28B, a cutoff would be chosen to separate the high-grade population; this cutoff could also be established from mining experience and milling considerations. For this study we retained the reference threshold value $z_0 = 1000$. That cutoff corresponds to the 0.97-quantile of the sample b-distribution, as opposed to being the 0.93-quantile of the reference Ely2 z-distribution. Blasthole indicator data defined at the cutoff value z_0 were retrieved, and their standardized semivariograms were calculated (see Fig. IV.28C). The anisotropic indicator semivariogram model fitted is:

$$\gamma_I^f(\mathbf{h}) = 0.1 + 0.9\text{Exp}\sqrt{\left(\frac{h_1}{60}\right)^2 + \left(\frac{h_2}{30}\right)^2} \qquad \text{(IV.13)}$$

where the superscript f denotes that this model was inferred from future blasthole data available on the 22×22 grid.

Ordinary indicator kriging is performed to evaluate at each grid node $\mathbf{u}$ the probability $p^{**}(\mathbf{u})$ that this node belongs to the low-grade mineralization:

$$p^{**}(\mathbf{u}) = \text{Prob}\{B(\mathbf{u}) \leq z_0 | b\text{-data}\} \qquad \text{(IV.14)}$$

Note that this probability refers to blasthole b-values, not to grade z-values. If mining experience had shown a bias between b-data and z-values, and that bias was assessed from, say, calibration data, the probability (IV.14) could be corrected appropriately. For this study, we assimilate $p^{**}(\mathbf{u})$ to the probability that the underlying true z-grade is below z_0.

The grade $z(\mathbf{u})$ itself could be evaluated in three steps:

1. Assuming that $z(\mathbf{u}) \leq z_0$, perform *ordinary* kriging using only the neighboring low-grade b-data $\{b(\mathbf{u}_\alpha^{(1)}) \leq z_0, \alpha = 1, \ldots, n_1(\mathbf{u})\}$. To this purpose, the variogram of the low-grade b-data was modeled as (see Fig. IV.28D):

$$\gamma_{Z_1}^f(\mathbf{h}) = 0.1 + 0.9\mathrm{Exp}\sqrt{\left(\frac{h_1}{80}\right)^2 + \left(\frac{h_2}{40}\right)^2} \qquad \text{(IV.15)}$$

   The kriging of the low-grade data provides the point support grade estimate:

$$z_1^{**}(\mathbf{u}) = z^*(\mathbf{u})|\mathbf{u} \in \text{low-grade population}$$

2. Similarly, perform *simple* kriging, using only the neighboring high-grade b-data $\{b(\mathbf{u}_\alpha^{(2)}) > z_0, \alpha = 1, \ldots, n_2(\mathbf{u})\}$ and the global high-grade sample mean 2584.65, to evaluate evaluate the point support z-grade, were it to belong to the second population:

$$z_2^{**}(\mathbf{u}) = z^*(\mathbf{u})|\mathbf{u} \in \text{high-grade population}$$

   For this study, simple kriging was used instead of ordinary kriging to palliate the sparsity of high-grade sample b-data: the local sample size $n_2(\mathbf{u})$ of high-grade data is small. Typically, the variogram of high-grade b-data is difficult to infer, again because of data sparsity. For this study, assuming that geological interpretation has recognized anisotropy of the high-grade population, we simply modified the previous low-grade variogram model (IV.15) by increasing the range anisotropy ratio:

$$\gamma_{Z_2}^f(\mathbf{h}) = 0.1 + 0.9\mathrm{Exp}\sqrt{\left(\frac{h_1}{100}\right)^2 + \left(\frac{h_2}{20}\right)^2} \qquad \text{(IV.16)}$$

3. The estimate of the point support z-grade is reconstructed as:

$$z^{**}(\mathbf{u}) = z_1^{**}(\mathbf{u})p^{**}(\mathbf{u}) + z_2^{**}(\mathbf{u})\left[1 - p^{**}(\mathbf{u})\right] \qquad \text{(IV.17)}$$

4. The $4 \times 4 = 16$ point support z-grade estimates discretizing each 11×11 SMU are then averaged into a final estimate $z_v^{**}(\mathbf{u})$ of the SMU grade. Actual selection is then performed on that estimated SMU grade [see relations (I.6–I.10)].

Figure IV.29 gives the grayscale maps of the kriging estimates $z_v^{**}(\mathbf{u})$ for the 11×11 and 22×22 SMU sizes and their statistics. When compared to the reference SMU grade maps of Figure IV.22C–E, the kriging maps and histograms display the characteristic smoothing effect: Some high-grade streaks were missed by kriging, especially on the 22×22 SMU size. The scattergrams of kriging estimates versus true values of Figures IV.29E and IV.29F indicate severe conditional bias; that is, overestimation of low SMU grades and underestimation of high SMU grades, yet the overall correlation between estimated and true SMU grades remains high at 0.81 for the 22×22 SMU size.

The estimation scores over Ely2, as displayed in Figure IV.29, are poorer than those obtained for the original Ely reference data set and displayed in Figure IV.14. There are two reasons for this: (i) the multipopulation character of Ely2 and its larger nugget effect, and (ii) the poorer quality of the blasthole data available over Ely2. Unfortunately for gold mines, Ely2 is more akin to a gold deposit with vein mineralization superimposed onto a pervasive lower-grade mineralization.

Results of actual selection: Ely2

Actual selection is performed on the SMU kriging estimated grades $z^{**}(\mathbf{u})$ displayed in Figures IV.29A and IV.29B. To provide an assessment of imperfect classification caused by using both sparse and imperfect blasthole data, the following maps are provided for selection of SMU of size 11×11 at cutoff grades $z = 139$ (Fig. IV.30, left column) and $z = 854$ (Fig. IV.30, right column):

- Figures IV.30A and IV.30B give the recovery under perfect selection, recalled from Figures IV.23B–E.
- Figures IV.30C and IV.30D give the actual recovery maps corresponding to actual selection.
- Figures IV.30E and IV.30F give the SMUs wrongly classified as ore (black pixels) and the SMUs wrongly classified as waste (gray pixels). The former correspond to misclassification area IV in the scattergram of Figure I.3, the latter to misclassification area III.

The actual recovery indicators of Figures IV.30C and IV.30D display a smoother spatial distribution when compared to the perfect recovery indicators for the same cutoff grade (see Figs. IV.30A and IV.30B). Note the complete lack of ore-declared SMUs in the northern part of Figure IV.30D, compared to scattered pockets of ore in Figure IV.30B. The proportion of SMUs wrongly classified as waste (false-negative gray pixels in Figs. IV.30E and IV.30F) increases with the cutoff grade. At the highest economic cutoff $z = 854$, such proportion dominates the proportion of SMUs wrongly classified as ore (false-positive black pixels).

The actual recovery indicators, such as those displayed in Figures IV.30C and IV.30D, provide the spatial distribution of actual tonnage recovered. The corresponding true SMU grades $z_v(\mathbf{u})$ are recovered, thus defining the actual recovered quantities of metal and mean ore grades [see expressions (I.8) and (I.9)]. Table IV.6 gives the actual recovery figures (tonnage, quantity of metal, mean ore grade, and profit) for the five cutoff values retained and the two

11 × 11 SMU				
Cutoff (z)	% Tonnage	% Q. of Metal	Mean Grade	Profit
76	69.1	98.4	523	305
139	45.7	88.4	710	440
271	20.7	65.0	1151	396
445	11.1	51.5	1705	372
854	3.8	30.4	2902	154
22 × 22 SMU				
Cutoff (z)	% Tonnage	% Q. of Metal	Mean Grade	Profit
76	72.2	98.8	502	354
139	46.2	87.5	696	481
271	21.3	63.9	1101	428
445	11.8	48.8	1512	366
854	3.6	29.1	3010	194

Table IV.6: Ely2 recovery values under actual selection from future blasthole data. % tonnage refers to total tons = 286 × 286 = 81796 unit tons (at zero cutoff), % quantity of metal refers to total = 81796 × 367 unit tons × grade unit (367 is the overall mean grade), and profit is given in $M\$$.

SMU sizes. Figure IV.31 gives a graphical representation of these recovery values as functions of the cutoff grade z.

Note the smaller profit for actual selection (Fig. IV.31D), as compared to that achieved under perfect selection (Fig. IV.25D). When compared to the profit achieved under actual recovery from the original Ely deposit (see Fig. IV.17), the profit under actual selection for Ely2 is much higher. This is primarily caused by the extreme high-grade values of Ely2, even though the Ely2 future blasthole data were of poorer quality than those used for the original Ely deposit.

The difference between the actual selection recovery characteristics (see Fig. IV.31 and Table IV.6) and the perfect selection recovery characteristics (see Fig. IV.25 and Table IV.4) quantifies the effect of imperfect selection. Figure IV.32 gives the relative difference between actual and perfect recovery values as functions of the cutoff grade z [see definition expression (IV.11)]. Recall that relative differences can be either negative (lower actual recovery) or positive (higher actual recovery), except for profit, where the difference is always negative (actual profit is necessarily less than profit under perfect selection).

As expected, the loss of profit caused by inaccurate selection is maximum for a highly selective operation (highest cutoff $z = 854$) for both SMU sizes (see Fig. IV.32D). The maximum profit loss is 65% for the 11 × 11 SMU size and 46% for the 22 × 22 SMU size. These figures indicate the critical importance of the information effect and the need to obtain good quality data at the time of actual selection.

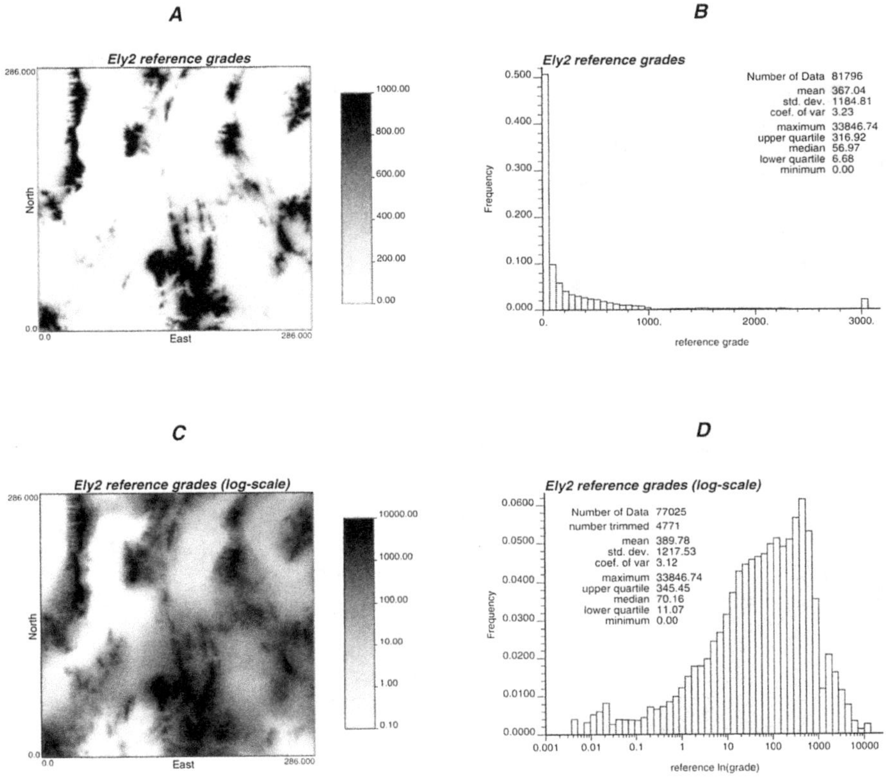

Figure IV.19: Grayscale plots (left) and histograms (right) of the Ely2 reference data set. **A, B:** arithmetic scale displaying vein-type structures of highest values, **C, D:** logarithmic scale displaying pervasive structures of low-to-median values. The 4771 zero values are trimmed out of the histogram and statistics of **D**.

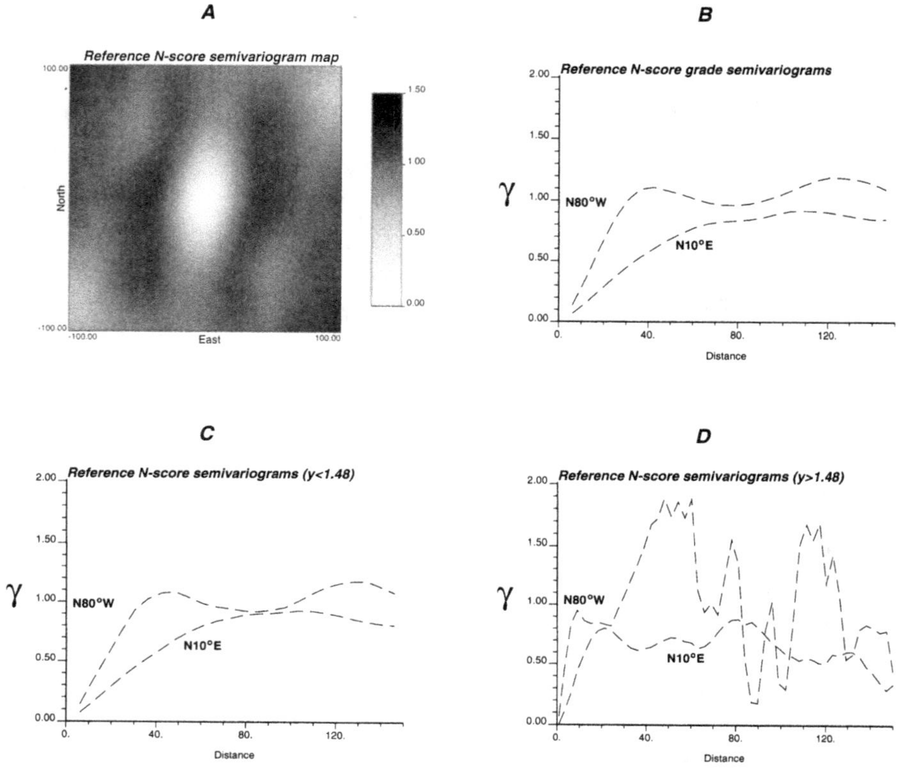

Figure IV.20: Ely2 reference semivariograms. **A:** normal scores semivariogram map, **B:** normal scores semivariograms in directions N10°E and N80°W, calculated across both populations, **C:** normal score semivariograms of the low-grade population (≤ 1.48), **D:** normal score semivariograms of the high-grade population ($y_0 = 1.48$ is the 93rd percentile of the standard normal distribution, which corresponds to the same percentile $z_0 = 1000$ of the reference Ely2 distribution).

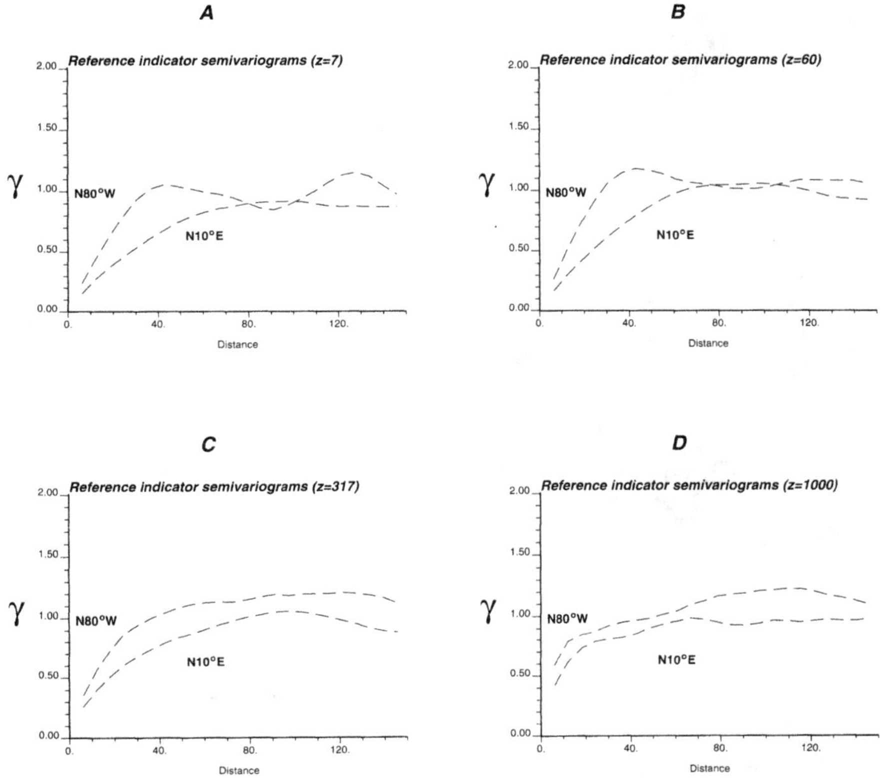

Figure IV.21: Ely2 reference indicator semivariograms along N10°E and N80°W directions. **A:** threshold $z = 7$ (lower quartile), **B:** threshold $z = 60$ (median), **C:** threshold $z = 317$ (upper quartile), **D:** threshold $z = 1000$ (separating the two populations).

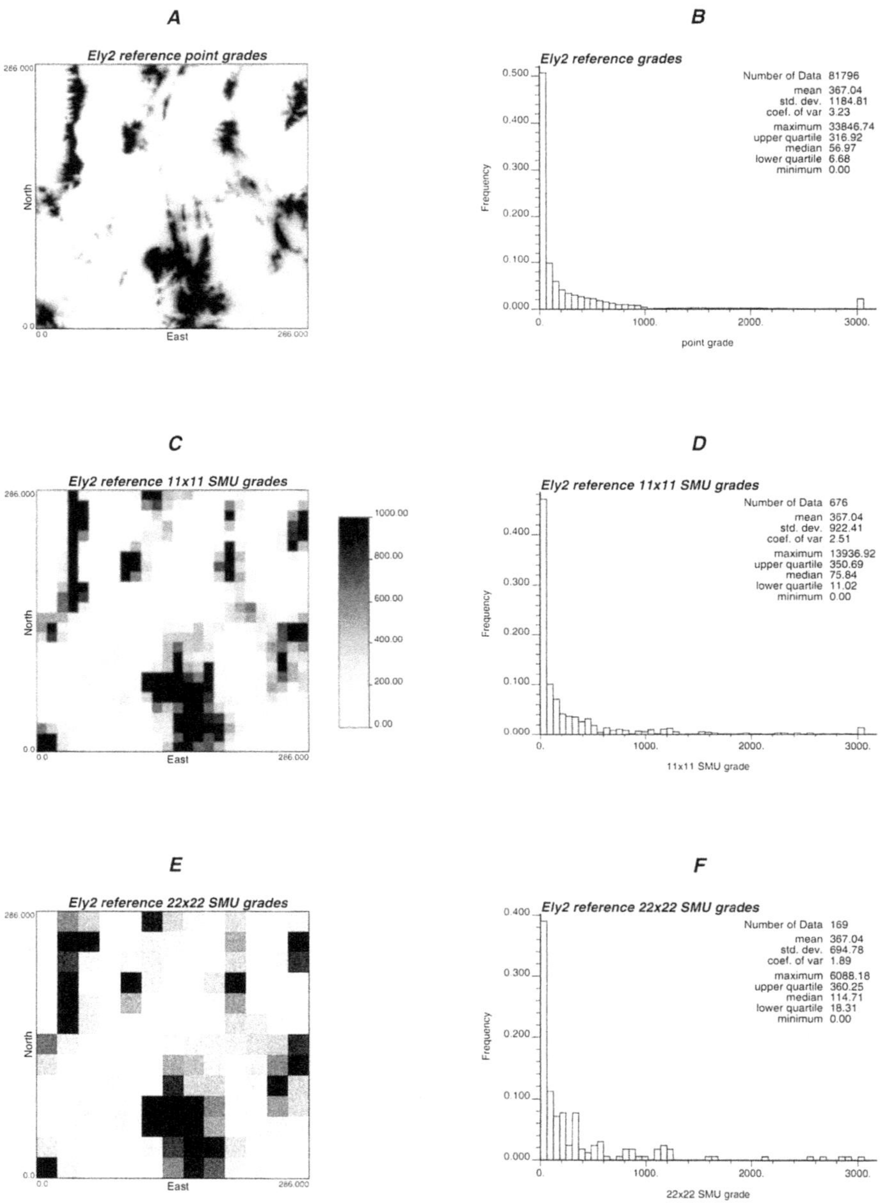

Figure IV.22: Grayscale plots (left) and histograms (right) of the Ely2 reference point and block support grades. **A**, **B:** point support grades and their histogram, **C**, **D:** 11×11 SMU grades and their histogram, **E**, **F:** 22×22 SMU grades and their histogram.

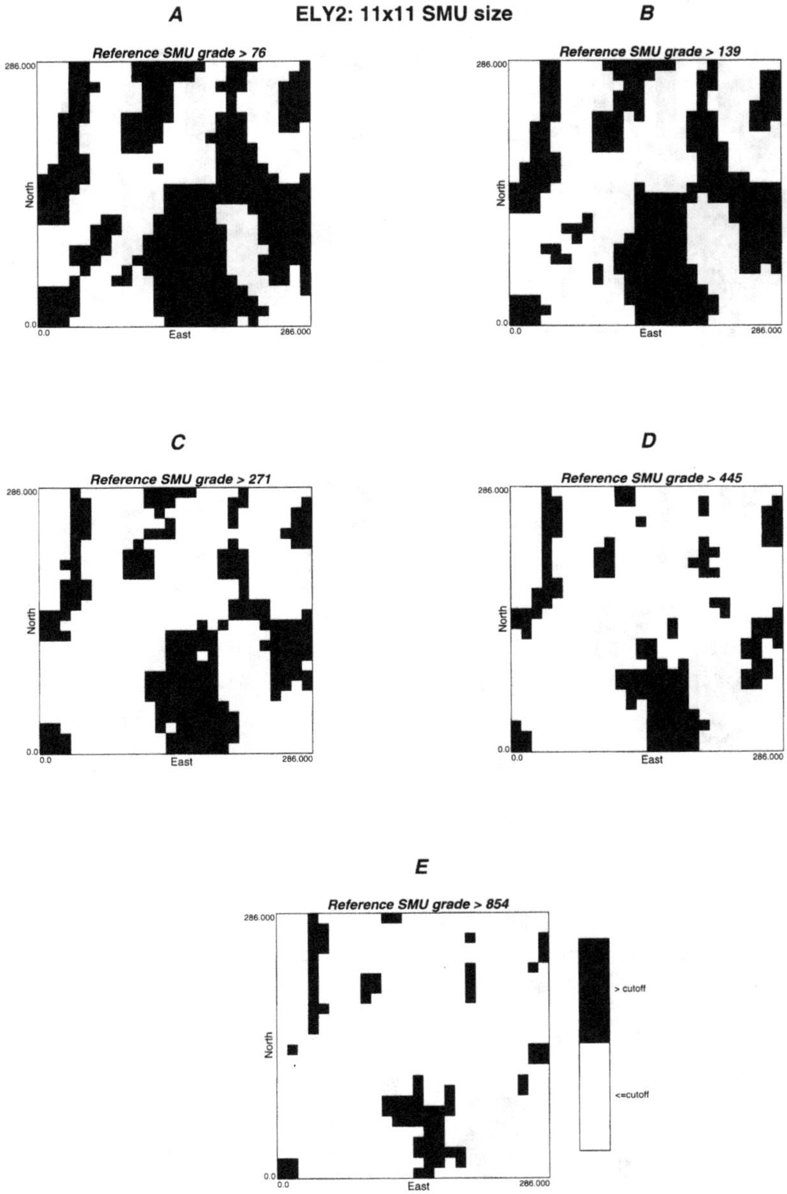

Figure IV.23: Ely2 maps of reference recovery indicators under perfect selection (SMU size 11 × 11).

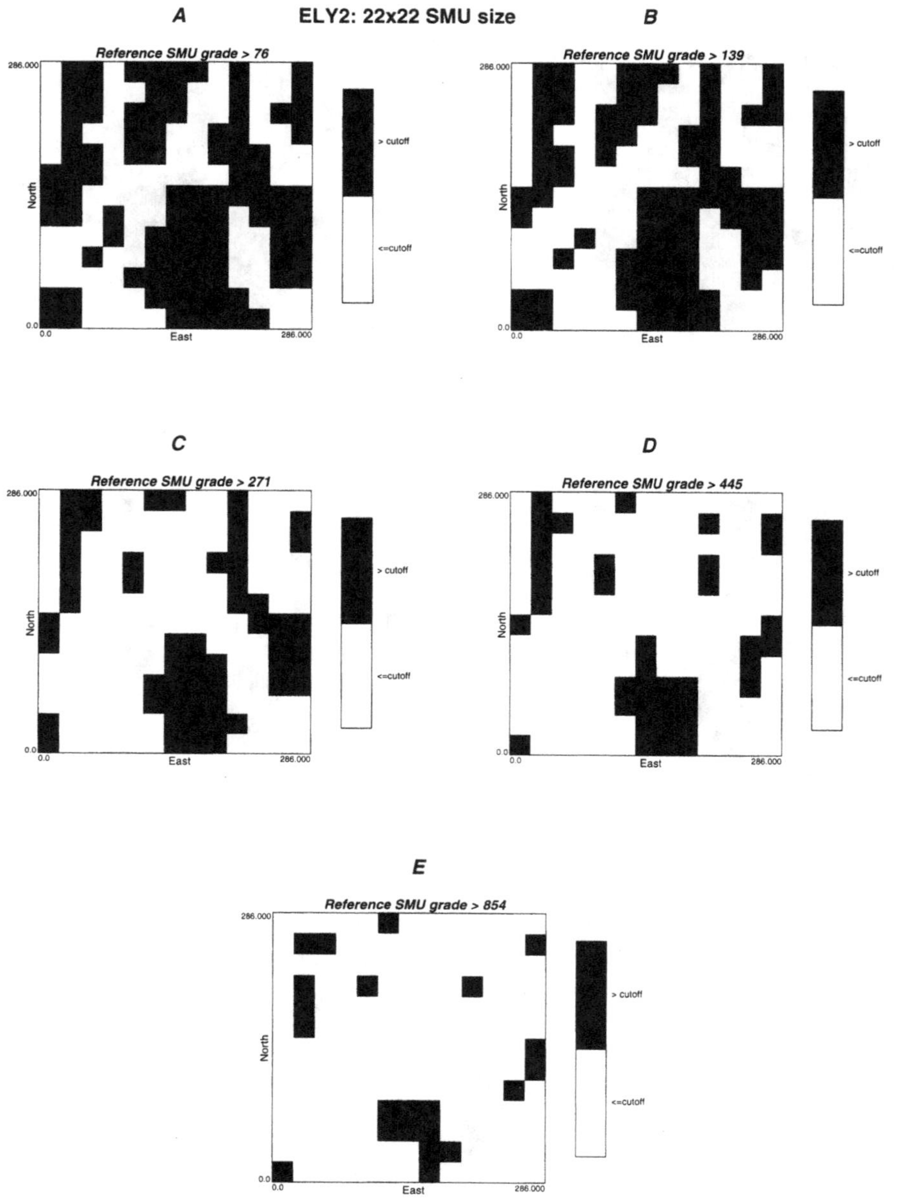

Figure IV.24: Ely2 maps of reference recovery indicators under perfect selection (SMU size 22×22).

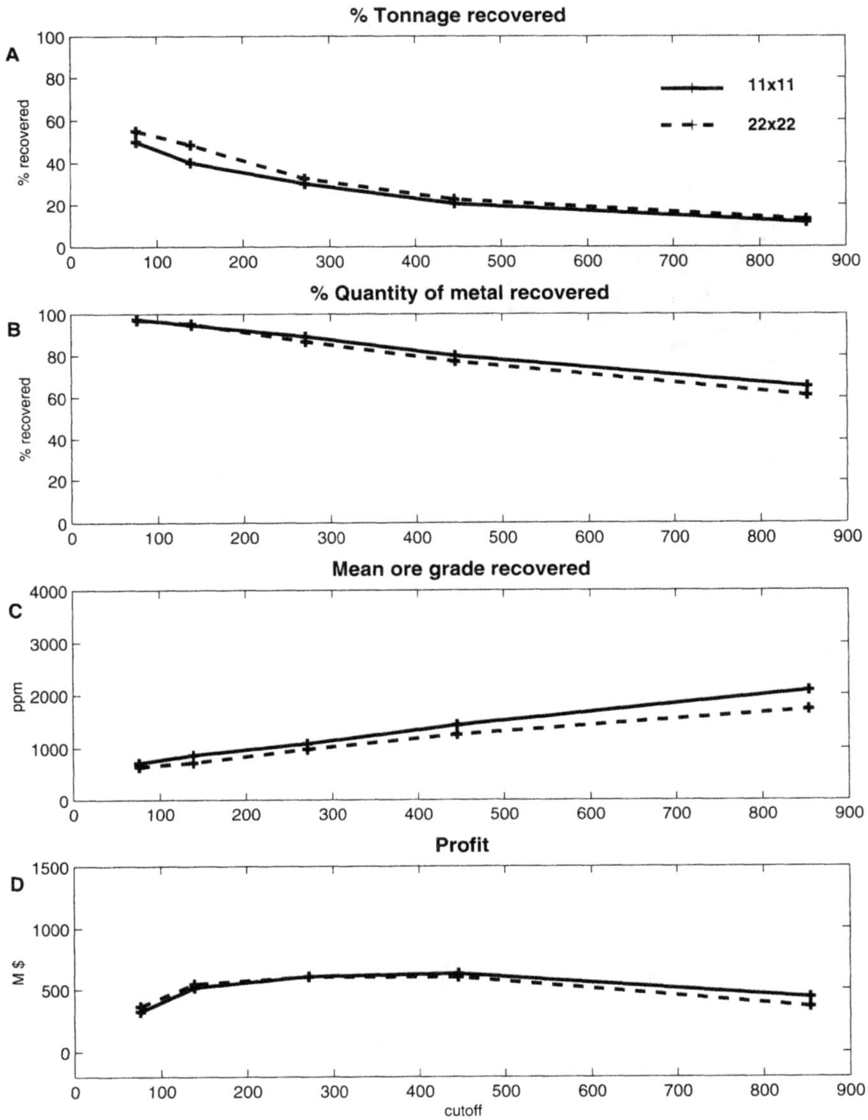

Figure IV.25: Ely2 recovery functions under perfect selection. **A:** percent tonnage recovered (total tonnage is for zero cutoff), **B:** percent quantity of metal recovered, **C:** mean ore grade, **D:** profit.

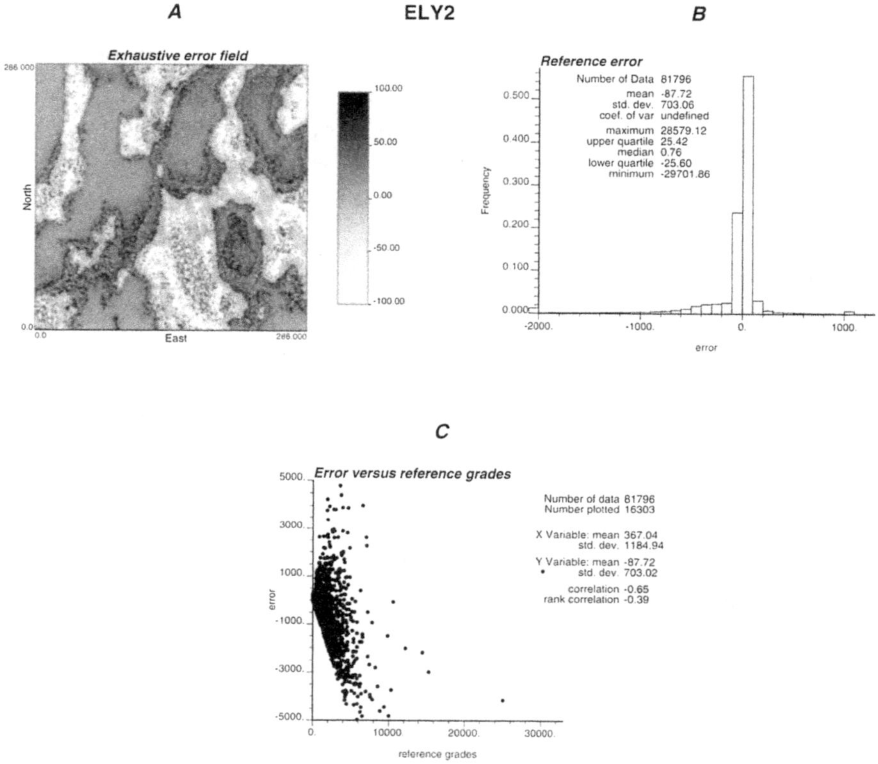

Figure IV.26: Ely2 statistics of the error r-field. **A:** pixel plot of the error r-values $[r(\mathbf{u}) = b(\mathbf{u}) - z(\mathbf{u})]$, **B:** histogram of r-values, **C:** scattergram of r- versus true z-values (note the error heteroscedasticity). Although only a few points are plotted, the statistics are calculated from all 81796 pairs.

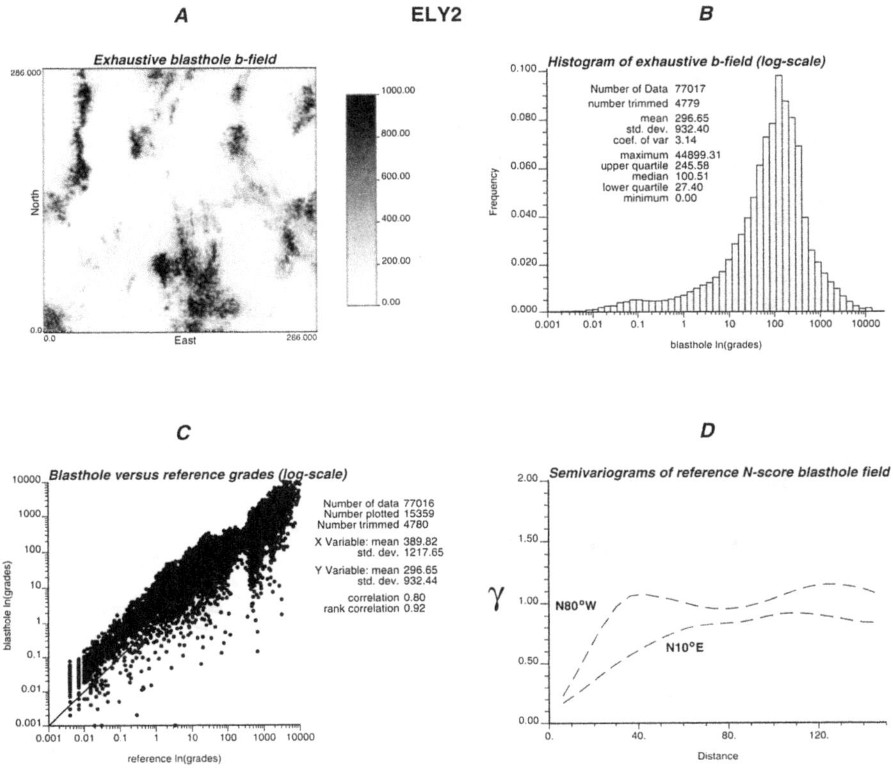

Figure IV.27: Ely2 statistics of the exhaustive blasthole b-field. **A:** grayscale plot of blasthole b-data, **B:** histogram of b-values (log-scale; all zero values are trimmed out), **C:** scattergram of b- versus reference z-values (log-scale), **D:** standardized b-semivariograms along N10°E and N80°W.

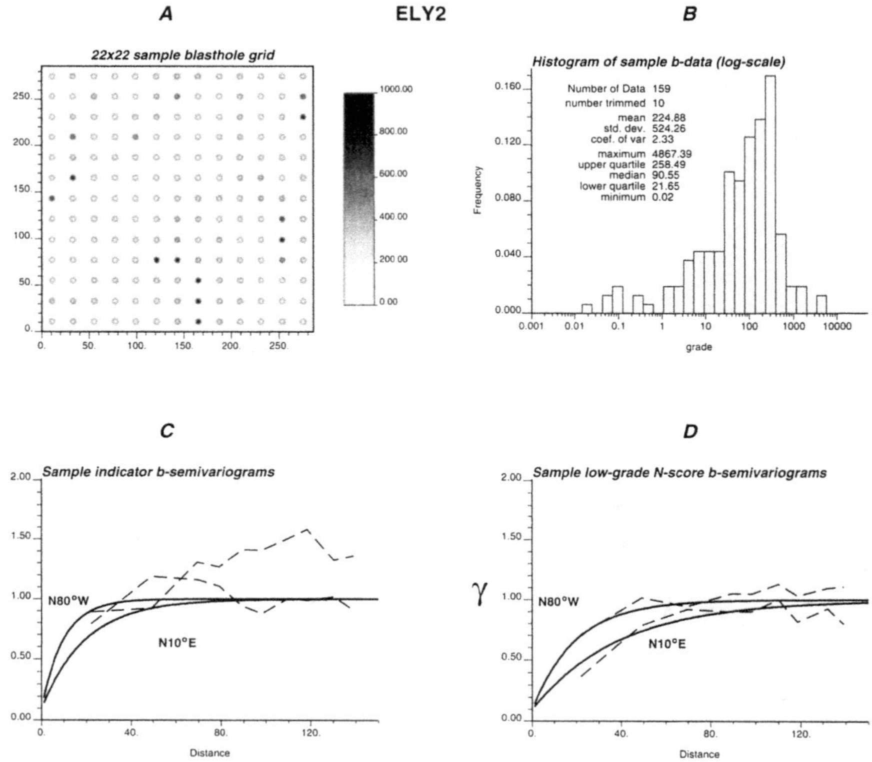

Figure IV.28: Ely2 future sample blasthole data on a 22 × 22 grid. **A:** location map of future *b*-samples, **B:** histogram of *b*-samples, **C:** standardized indicator semivariograms along directions N10°E, N80°W, and model fit, **D:** standardized low-grade *b*-semivariograms and model fit.

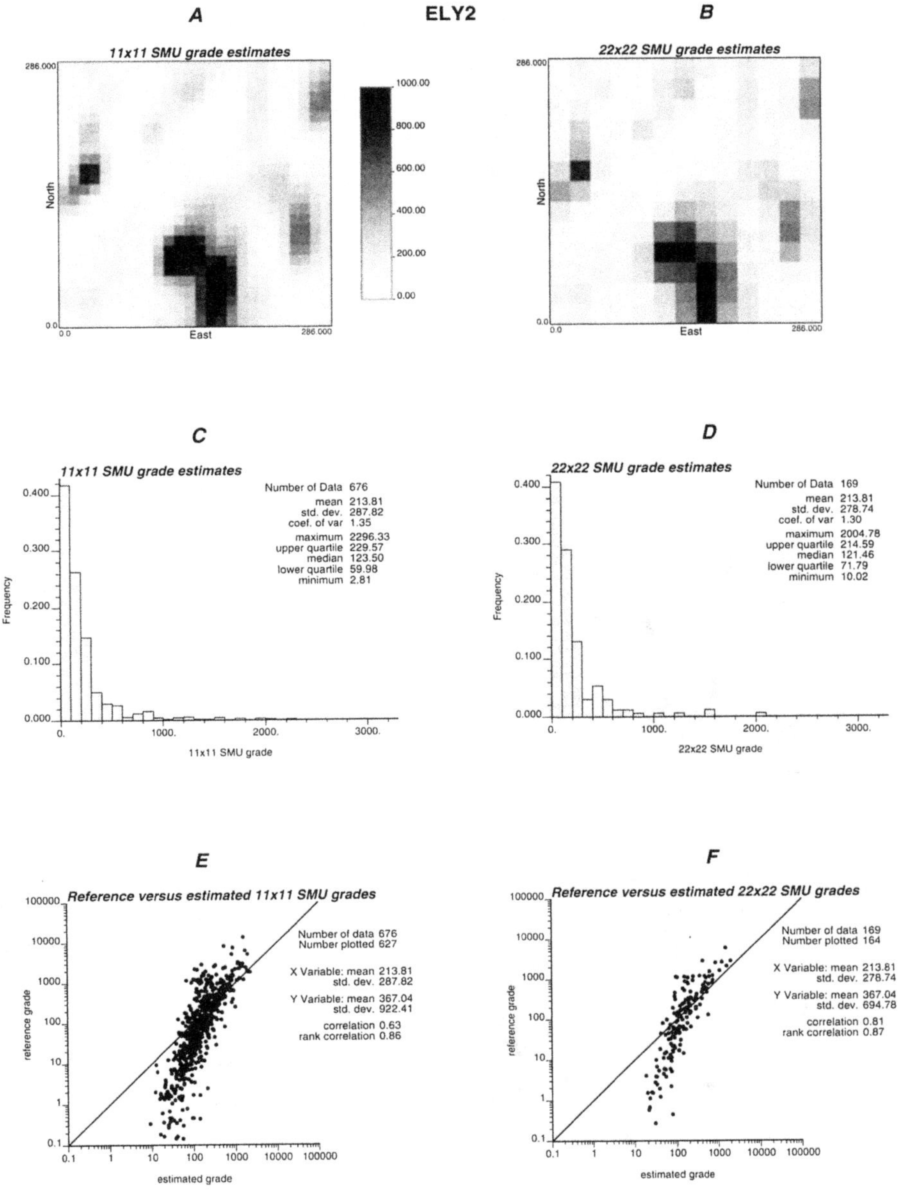

Figure IV.29: Statistics of Ely2 block support grade estimates of SMU size 11×11 (left) and 22×22 (right), obtained via ordinary kriging using future blasthole data. **A**, **B:** grayscale plots, **C**, **D:** histograms, **E**, **F:** scattergrams of true versus estimated grades. Zero values are not plotted in the scattergrams, but were accounted for in the statistics.

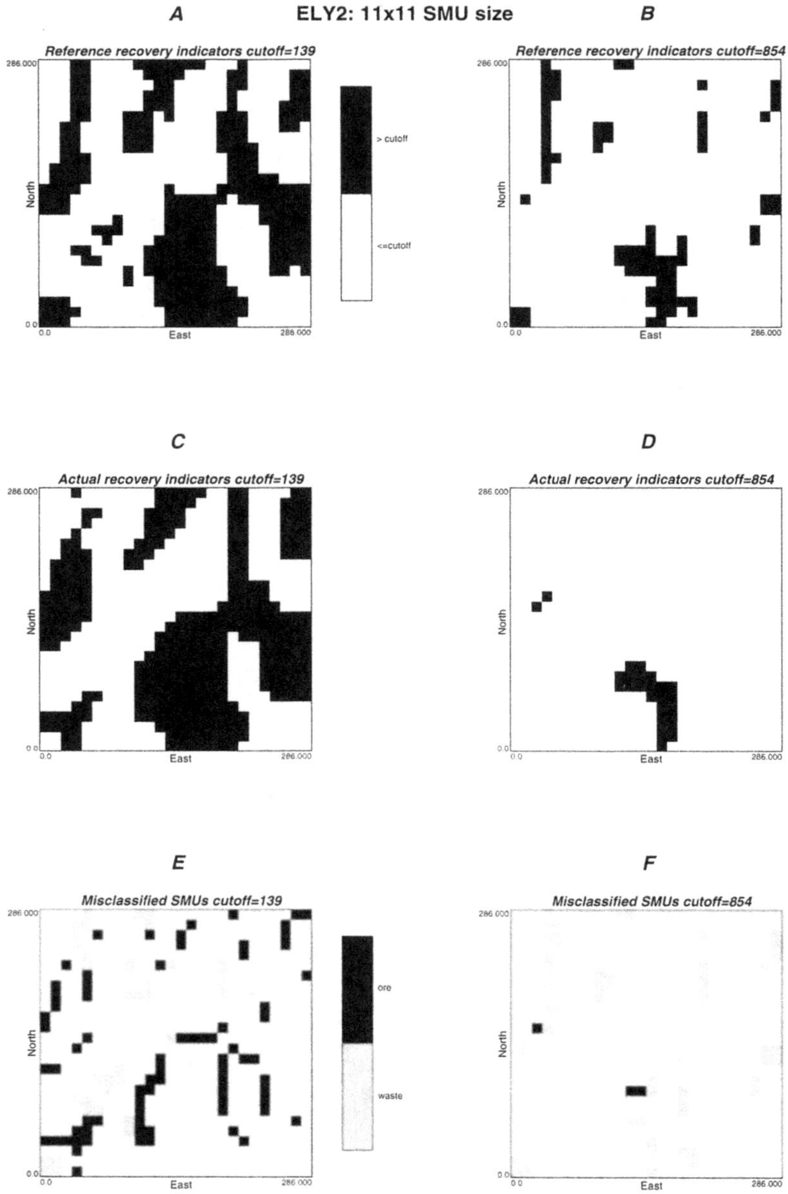

Figure IV.30: Classification scores of 11×11 SMU grades estimated using future blasthole data for cutoff $z = 139$ (left) and $z = 854$ (right). **A**, **B:** reference recovery indicators under perfect selection, **C**, **D:** reference recovery indicators under actual selection, **E**, **F:** misclassified SMUs: as ore (black pixels) and as waste (gray pixels).

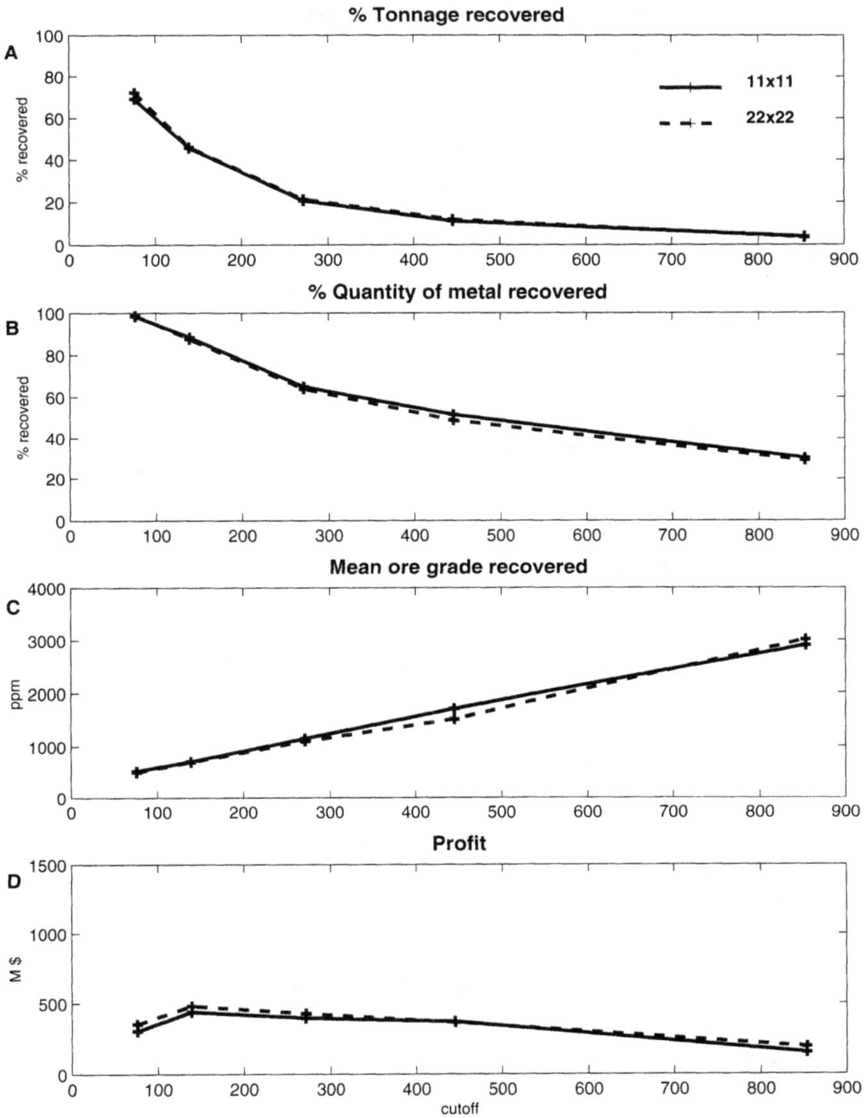

Figure IV.31: Ely2 recovery functions under actual selection using future blasthole data. **A:** percent tonnage recovered (total tonnage is for zero cutoff), **B:** percent quantity of metal recovered, **C:** mean ore grade, **D:** profit.

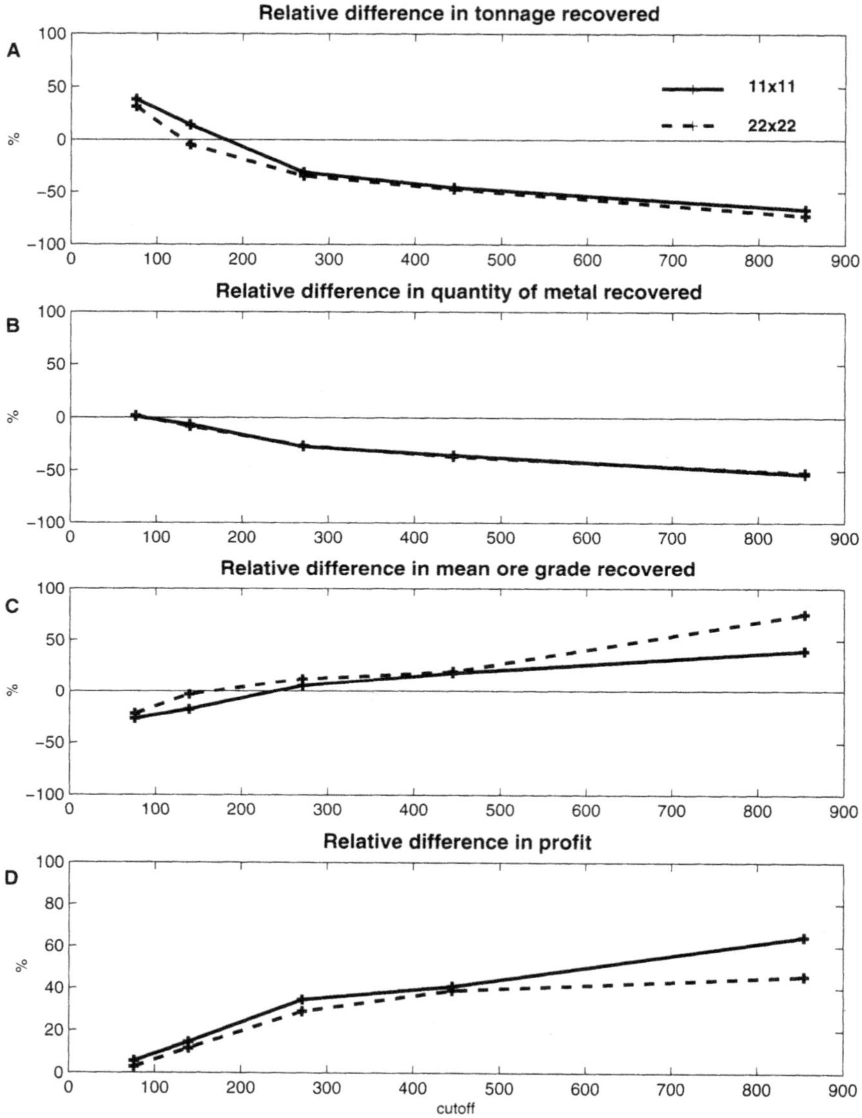

Figure IV.32: Ely2 relative difference in recovery functions under actual and perfect selection (profit under actual selection is always smaller than under perfect selection, and hence the absolute relative difference is given).

IV.5 Summary

Although the foundation of any estimation or simulation algorithm has to be established from theory, it is essential to check its performance and robustness on real data sets. Any theory starts with simplifying assumptions about the structures and statistics of the unknowns. Real data, particularly earth sciences data, always defy such assumptions.

To check spatially distributed estimates one should have available an exhaustive data set that could provide the reference values but also the estimated values based on a sample of these true values. Because no mining operation can ever provide the true SMU grade values over an area/panel large enough to alleviate statistical fluctuations, we retained a topographic data set from the Ely West United States Geological Survey quadrangle in Nevada. This "natural" data set was transformed to match a histogram and statistics more akin to mineral grades; the grade unit has been deliberately left undetermined. The transform left unchanged the spatial structures of the original Ely elevation values. This first data set proved to be too congenial, displaying one single grade population with Gaussian-type high entropy (homogeneously heterogeneous). Therefore, a second data set, Ely2, was generated from Ely, such as to display a second high-grade population with clear structured anisotropy as compared to the bulk low-to-median-grade population. The proposed algorithms for evaluation of recoverable reserves will be tested successively on both reference data sets, Ely and Ely2, with results expected to be poorer over Ely2.

The reference (true) recovery functions, to be estimated or later simulated, consist of four functions of the cutoff grade, tonnage, quantity of metal, ore grade, and profit. Of course these recovery functions depend on the size of the SMU.

There are two sets of recovery functions:

- The first corresponds to *perfect* selection, ignoring any misclassification: each SMU is selected as ore or waste on the basis of its true grade $z_v(\mathbf{u})$.

- The second set corresponds to *actual* selection, accounting for misclassification: selection of any SMU as ore is based on its ultimate estimated value $z_v^{**}(\mathbf{u})$ at the time of mining. The difference between $z_v(\mathbf{u})$ and $z_v^{**}(\mathbf{u})$ is the source of misclassification.

Determination of the reference actual recovery calls for knowing, in addition to the true values $z_v(\mathbf{u})$, the future data from which the future selection estimates $z_v^{**}(\mathbf{u})$ will be built. Thus, for each of the two reference data sets, Ely and Ely2, we had to simulate these future (blasthole) data and build from them the reference future selection estimates $z_v^{**}(\mathbf{u})$. The pairs $\{z_v(\mathbf{u}), z_v^{**}(\mathbf{u})\}$ represent the reference data sets from which the reference (true) actual recovery functions are determined. This actual recovery is later evaluated (in Section V.3) through simulated pairs, then denoted $\{z_v^{(l)}(\mathbf{u}), z_v^{**(l)}(\mathbf{u})\}$ for the lth realization.

At this point, the reader should be comfortable with the difference between a reference estimated value $z_v^{**}(\mathbf{u})$ and a simulated estimated value $z_v^{**(l)}(\mathbf{u})$; if not, we suggest a review of Section I.1 for notations and Section II.3 for the concept of simulation of selection estimates.

Chapter V

Simulation of Recovery: Application

In this chapter, the simulation approach proposed in Chapter II is applied to the evaluation of reserves associated with the two reference data sets introduced in Chapter IV. Only a sparse set of sample grades is retained for the task of evaluating recoverable reserves (30 samples out of the reference total of $286 \times 286 = 81796$ values). The missing information is simulated instead of being estimated; the difference between the many possible simulated grade realizations provides a measure of uncertainty regarding the spatial distribution of recoverable reserves.

Simulation is performed on the quasi-point support of the sample data available (Section V.1); assessment of uncertainty about the in situ mean (before any selection) is also addressed in this section. The simulated point grade values are then averaged into simulated "true" SMU grades: This allows assessment of the recovery functions under perfect selection (Section V.2). The future selection data can also be simulated; from these future data, simulated SMU "estimates" are derived: This allows simulation of the actual recovery process under imperfect selection (Section V.3). A sensitivity analysis of the results obtained to the variogram model adopted is also provided in this section. The last section deals with the second reference data set.

V.1 Simulation of Point Support Grades

As discussed in the methodological Section II.1, any consistent simulation of SMU support average grades and their estimates starts with the simulation of point support grade values. These simulated point grades can be averaged into simulated grades of SMUs of any size, or combined into estimated SMU grades.

All simulated realizations must be conditioned to the information available at the (present) time of the prediction study. That information includes:

- the original sample data, here taken of size $n = 30$, and
- any soft information, such as structural information related to the variogram model(s).

A sparse set of 30 samples $\{z(\mathbf{u}_\alpha), \alpha = 1, \ldots, n = 30\}$ was retained from the map of the Ely reference grades (see sample locations in Fig. V.1A). Thirty drillholes may be too few to mimic the data available at the time of detailed mine planning, but would be a reasonable number for an assessment phase. Random stratified sampling was considered to avoid any bias (see the quantile-quantile plot of Fig. V.1C, comparing the sample histogram of Fig. V.1B to the reference histogram of Fig. IV.1B). Note the lack of samples in the NE corner: This was deliberate to allow checking the estimation/prediction algorithms in extrapolation situations. Note also the failure of a sparse sample in evaluating accurately the statistics of extreme values (above $z = 500$), and the corresponding underestimation of the global variance (the sample coefficient of variation is 1.57, whereas the reference value is 1.82). Such inaccurate evaluation of extreme values from early sparse sample data is unfortunately common (e.g., Koch and Link, 1970–71, vol. 1, p. 70). Figure V.1D gives four directional semivariograms calculated from the sample $n = 30$ normal score data. Again, not atypical, these sample semivariograms do not succeed in revealing the actual anisotropy of the reference z-field (compare to Figs. IV.2B and IV.2C). In practice, geological interpretation may be able to establish the existence of anisotropy, and that information might be introduced in the semivariogram model adopted.

In the following, we will consider two scenarios to account for uncertainty about the semivariogram model used in generating the simulated deposits $\{z^{(l)}(\mathbf{u}), \mathbf{u} \in D\}$, $l = 1, \ldots, L$:

- **Scenario 1:** the exhaustive reference normal score semivariogram model (IV.1) is retained. Note that, as kriging variances are not used in the selection process, the total sill of the semivariogram model is irrelevant.
- **Scenario 2:** an isotropic semivariogram model is retained, with range inferred from the sample semivariogram of Figure V.1D:

$$\gamma_Y(\mathbf{h}) = \gamma_Z(\mathbf{h}) = \text{Exp}\left(\frac{|\mathbf{h}|}{80}\right) \tag{V.1}$$

Scenario 1 corresponds to the best case, and scenario 2 to a worst case, ignoring any geological information about anisotropy.

Under each of the two semivariogram scenarios, a set of $L = 50$ realizations of point support grade z-values were generated using the sequential Gaussian simulation algorithm and GSLIB program `sgsim` (Deutsch and Journel, 1998,

p. 170). These realizations were conditioned to the $n = 30$ sample data available and were generated over the reference panel V of size 286×286.

Figure V.2 gives the `sgsim` parameter file used to generate the 50 grade realizations $\{z^{(l)}(\mathbf{u}), \mathbf{u} \in D\}$ under scenario 1. These realizations were ranked according to their overall mean values, and Figure V.3 gives the grayscale maps and histograms of the realizations with highest mean (400), median mean (257), and lowest mean (156). The sample histogram and the anisotropy of the reference normal score semivariogram model (IV.1) adopted is reflected on all realizations within statistical fluctuations (see Fig. V.4). As expected, a mere two-point statistic such as a semivariogram fails to reproduce the reference (unknown) complex pattern of heterogeneities (compare Figs. V.3 and IV.1A). For discussions on two-point versus multiple-point statistics, and the limitation of the semivariogram as a structural analysis tool, refer to Journel (1993), Guardiano and Srivastava (1993), Caers and Journel (1998), and Strebelle (2002).

Figures V.5 and V.6 give the grayscale maps and histograms of the simulated true SMU grades (for 11×11 and 22×22 block size), corresponding to the three realizations of Figure V.3. Compare the simulated realizations of Figures V.3, V.5, and V.6 to the reference maps of Figure IV.5. Interestingly, if simulation from a mere semivariogram model fails to reproduce the complex spatial structure of the reference point support grade (Fig. IV.5A), it does a fairly good job in reproducing the much blurred structure at the SMU support (Figs. IV.5C–E). Indeed, because of volume averaging, it is easier to simulate block grades than point support grades.

Figures V.3, V.5, and V.6 indicate that the simulated realizations differ most in the NE area, where there are no sample grade data. Simulation fills in that area with the same texture generated elsewhere, but the simulated values can be high (V.3A) or low (V.3C) because there are no conditioning data in that area.

Mean Grade Uncertainty Assessment

Within a panel V of known geometry, and hence known tonnage, and assuming full recovery (zero cutoff grade), the only uncertainty relates to the unknown mean grade value. The distribution of the L simulated mean values over V provides an assessment of uncertainty about that unknown mean grade. A different approach is provided by the bootstrap technique (Efron, 1982), whereby multiple samples of equal size (here $n = 30$) are obtained by resampling with replacement the original (and only available) sample histogram.

Figures V.7A and V.7B give the histograms of the overall mean of the $L = 50$ simulated point support grade realizations under semivariogram scenarios 1 and 2. The ranges of uncertainty $[156, 400]$ and $[162, 400]$ are similar and both very large. This is not surprising for such a small sample size ($n = 30$) and such a skewed distribution.

The measure of uncertainty about the global mean given by the distributions of Figures V.7A and V.7B is entirely dependent on the ergodic properties of the $L = 50$ normal score realizations $\{y^{(l)}(\mathbf{u}), \mathbf{u} \in D\}$, $l = 1, \ldots, L$, before back transform to the sample grade z-histogram (Deutsch and Journel, 1998, p. 129). Were the panel V very large with regard to the largest range of the semivariogram model, the $y^{(l)}$-realizations would be exactly standard normal; their back transform would then restitute the z-sample mean (266) exactly; that is, with a zero uncertainty range. However, one would not expect perfect estimation of the mean from such a small sample size, even if the panel size is large.

Modeling uncertainty about global (nonrepetitive) parameters is a difficult task (Journel and Huijbregts, 1978, p. 410). Assessing such uncertainty through the ergodic fluctuations of realizations of a random function model, as done in Figures V.7A and V.7B, is indeed not fully satisfactory. A more direct and apparently better approach is to bootstrap the global mean from the sample histogram of Figure V.1B (e.g., Efron and Tibshirani, 1993). One would draw with replacement from the sample histogram, say 50 sets of 30 z-values each; retrieve from each set the sample mean; and build the histogram of these 50 resampled means (see results in Fig. V.7C). The range of uncertainty for the global mean is now $[148, 424]$, larger than that ($[156, 400]$) provided by Figure V.7A based on ergodic fluctuations. The problem with this classical bootstrap is that it requires that the original samples (here $n = 30$) be independent one from another, a hypothesis reasonable only in early development stages, when the drillholes are few and far apart without any preferential clusters (approximately the case here; see Fig. V.1A).

We have the luxury of knowing the underlying true reference values, and hence we can resample from that reference data set instead of the sample histogram. Again, 50 independent sets of 30 z-samples were drawn from the reference data set and the histogram of Figure IV.1B, and then the 50 sample means were calculated; their histogram is given in Figure V.7D, yielding $[151, 434]$ for the uncertainty interval. Notwithstanding the previous limitations of both the simulation and the bootstrap approach, they perform reasonably well in predicting the reference 50% probability interval $[211, 306]$ bounded by the lower and upper quartile values of the histogram of Figure V.7D; these predictions are:

- $[216, 310]$ for the simulation approach under variogram scenario 2; that is, the sample variogram, and
- $[238, 325]$ for the bootstrap approach.

The 95% probability intervals for the global mean are very large, no matter the approach used to evaluate them. Recall, though, that this mean relates to a panel V not a whole deposit: the larger V, the more data, the more error averaging.

The importance of good luck

Any reserves evaluation calls for sample data representative of the underlying grade population from which they were extracted, a property known as unbiasedness. No statistics nor any empirical approach can correct for any undetected data bias. If the data are erroneous, biased either by poor technology or ill will, and this bias is not detected and corrected, statistical processing assuming unbiasedness remains futile.

Unfortunately, the wide dispersion of the histograms of Figure V.7 show another source of uncontrollable error: bad luck. Even if the sampling is fair and evenly distributed over the domain, the sample mean may deviate severely from the true population mean. From Figure V.7D, the sample mean can be as low as 151, and with worst consequences, it can be as high as 434, a value far above the true mean of 265 grade units.

Had we retained the worst sample with mean 434, all our calculations would have been hopelessly biased, no matter the (reasonable) algorithm retained. This is the reason why throughout the demonstrations developed in this book, the sparse samples ($n = 30$) retained as actual original data were always "good luck" samples with a sample mean not too far away from the true mean; for example, for Ely the sample mean is 266 (see Fig. V.1B)—very close to the true mean of 265 grade units (see Fig. IV.1B).

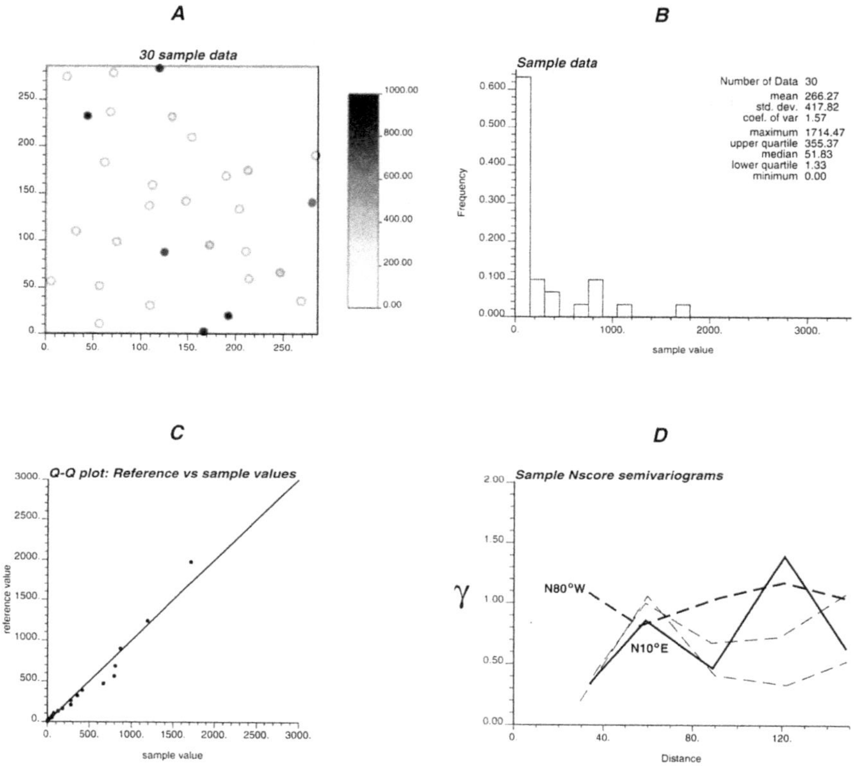

Figure V.1: Ely sample locations and statistics.

```
                         Parameters for SGSIM
                         ********************

START OF PARAMETERS:
sample.dat                      -file with data
1  2  0  3  0  0                -  columns for X,Y,Z,vr,wt,sec.var.
-100.0        1.0e21            -  trimming limits
1                               -transform the data (0=no, 1=yes)
sgsim.trn                       -  file for output trans table
0                               -  consider ref. dist (0=no, 1=yes)
nohistsmth.out                  -  file with ref. dist distribution
1  2                            -  columns for vr and wt
0.0   7000.0                    -  zmin,zmax(tail extrapolation)
1        1.5                    -  lower tail option, parameter
4        2.0                    -  upper tail option, parameter
0                               -debugging level: 0,1,2,3
sgsim.dbg                       -file for debugging output
sgsim50.out                     -file for simulation output
50                              -number of realizations to generate
286   0.5    1.0                -nx,xmn,xsiz
286   0.5    1.0                -ny,ymn,ysiz
1     0.5    1.0                -nz,zmn,zsiz
169077                          -random number seed
0    24                         -min and max original data for sim
12                              -number of simulated nodes to use
1                               -assign data to nodes (0=no, 1=yes)
1      3                        -multiple grid search (0=no, 1=yes),num
0                               -maximum data per octant (0=not used)
300.0 300.0  10.0               -maximum search radii (hmax,hmin,vert)
  0.0   0.0   0.0               -angles for search ellipsoid
0      0.60   1.0               -ktype: 0=SK,1=OK,2=LVM,3=EXDR,4=COLC
nodata.dat                      -  file with LVM, EXDR, or COLC variable
4                               -  column for secondary variable
2     0.0                       -nst, nugget effect
3     0.65 10.0    0.0    0.0   -it,cc,ang1,ang2,ang3
           60.0   35.0   10.0   -a_hmax, a_hmin, a_vert
2     0.35 10.0    0.0    0.0   -it,cc,ang1,ang2,ang3
          230.0   37.0   10.0   -a_hmax, a_hmin, a_vert
```

Figure V.2: **sgsim** parameter file for generating 50 simulated point support grade realizations under semivariogram scenario 1 (reference model).

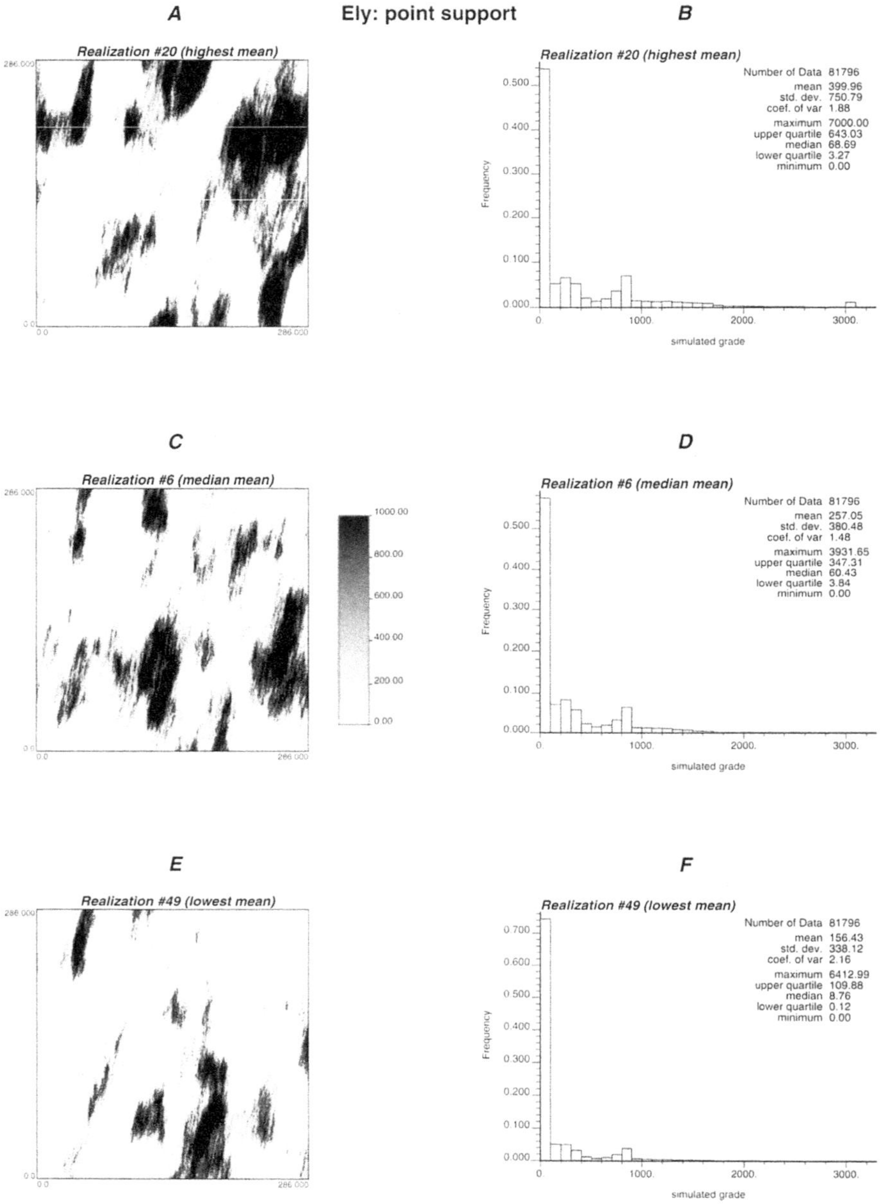

Figure V.3: Three simulated Ely point support grade realizations and their statistics under semivariogram scenario 1 (reference model).

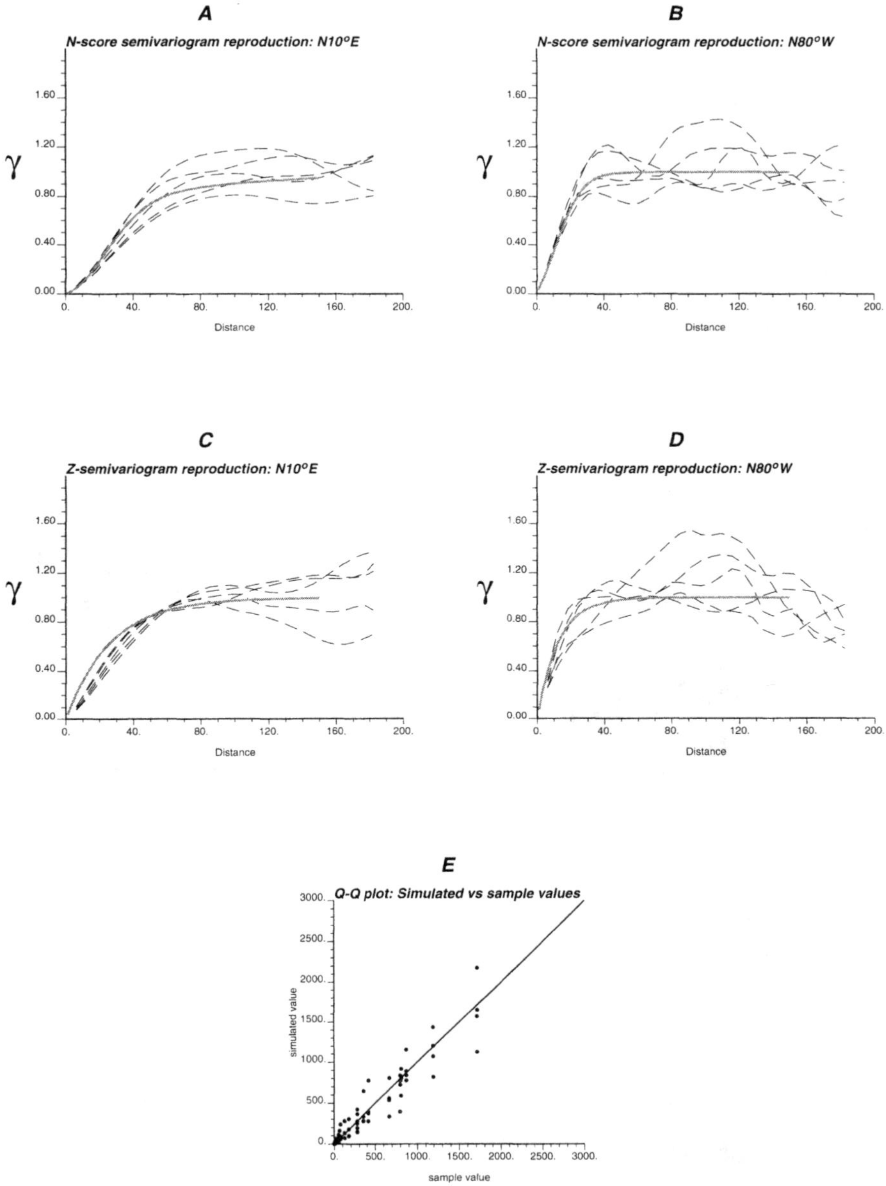

Figure V.4: Reproduction of sample statistics by the five first simulated realizations. **A**, **B:** normal score semivariogram reproduction, **C**, **D:** z-semivariogram reproduction (solid line: reference model, dashed lines: semivariogram of simulated values), **E:** quantile-quantile plot between simulated and sample distributions for the first realization.

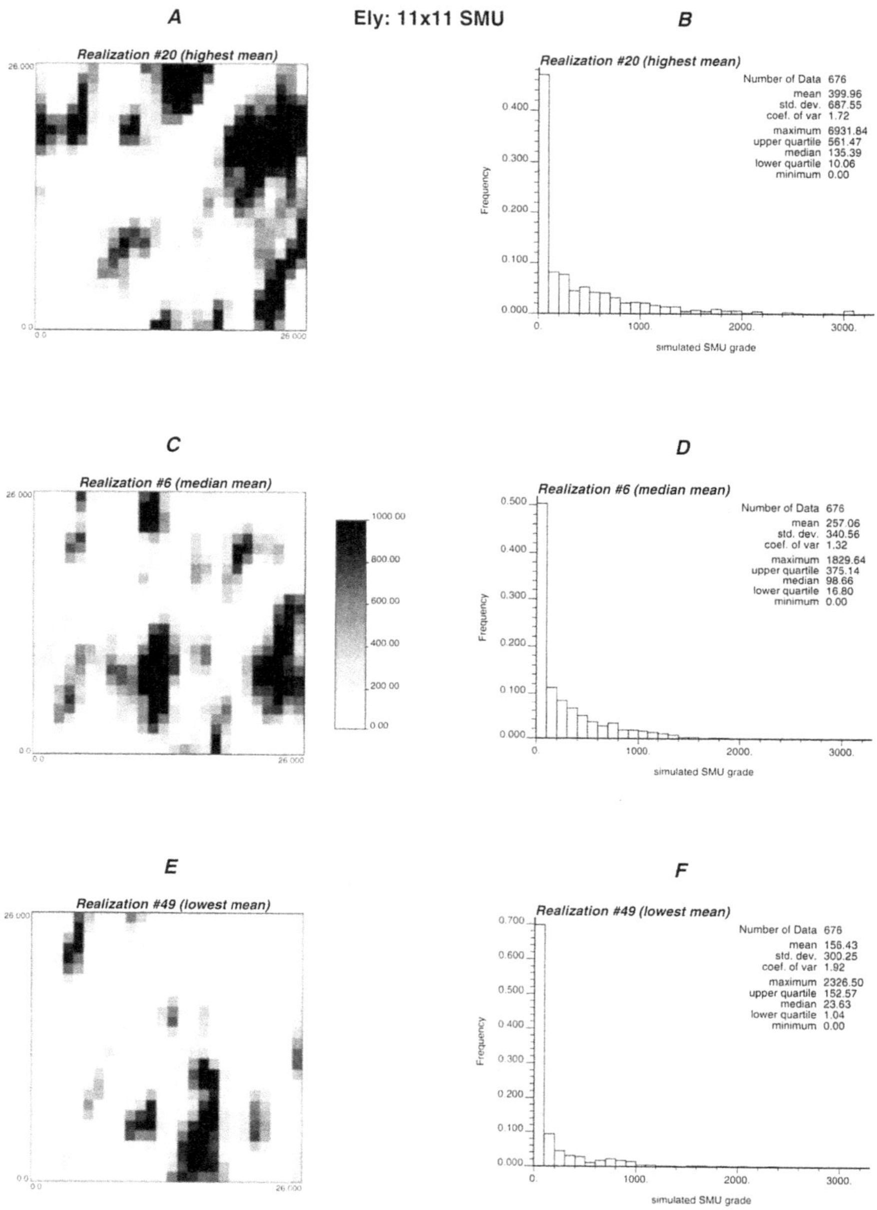

Figure V.5: Three simulated 11 × 11 SMU grade realizations and their statistics under semivariogram scenario 1 (reference model).

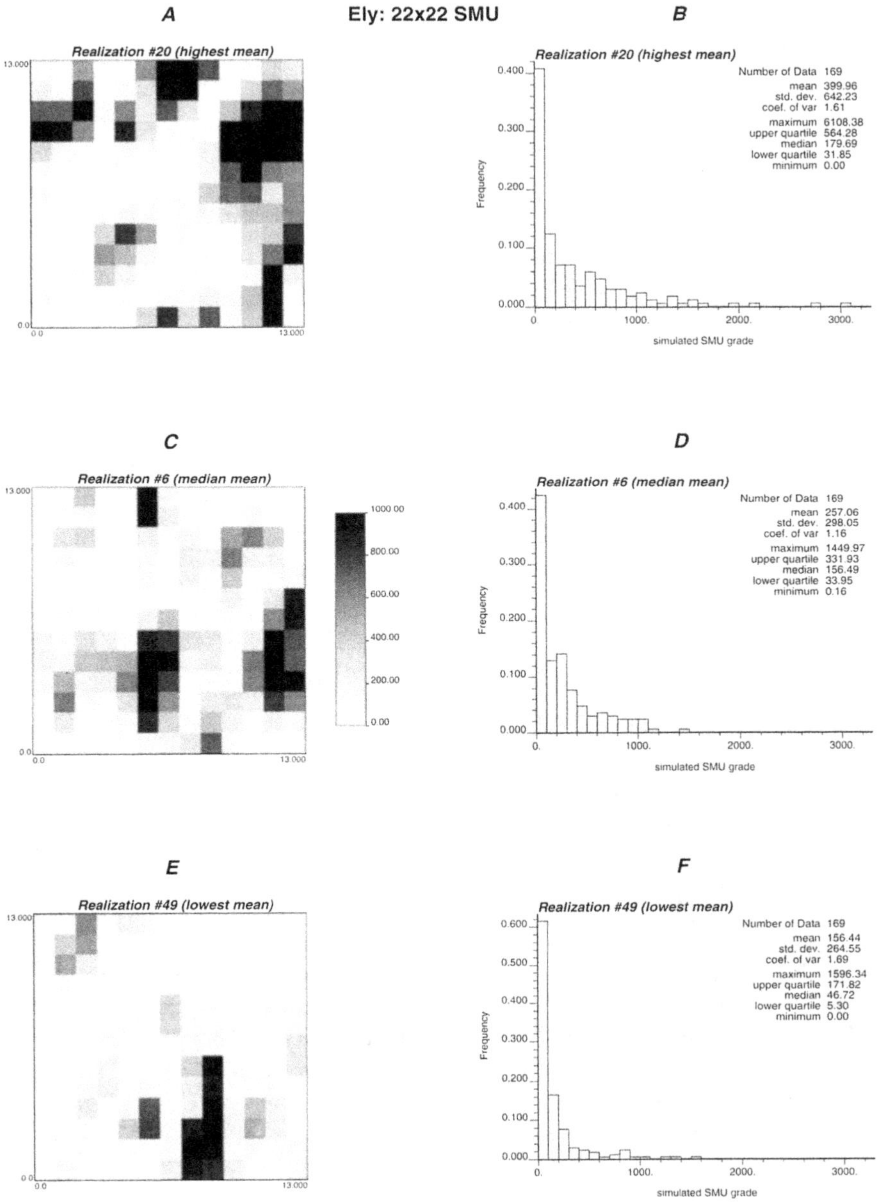

Figure V.6: Three simulated 22×22 SMU grade realizations and their statistics under semivariogram scenario 1 (reference model).

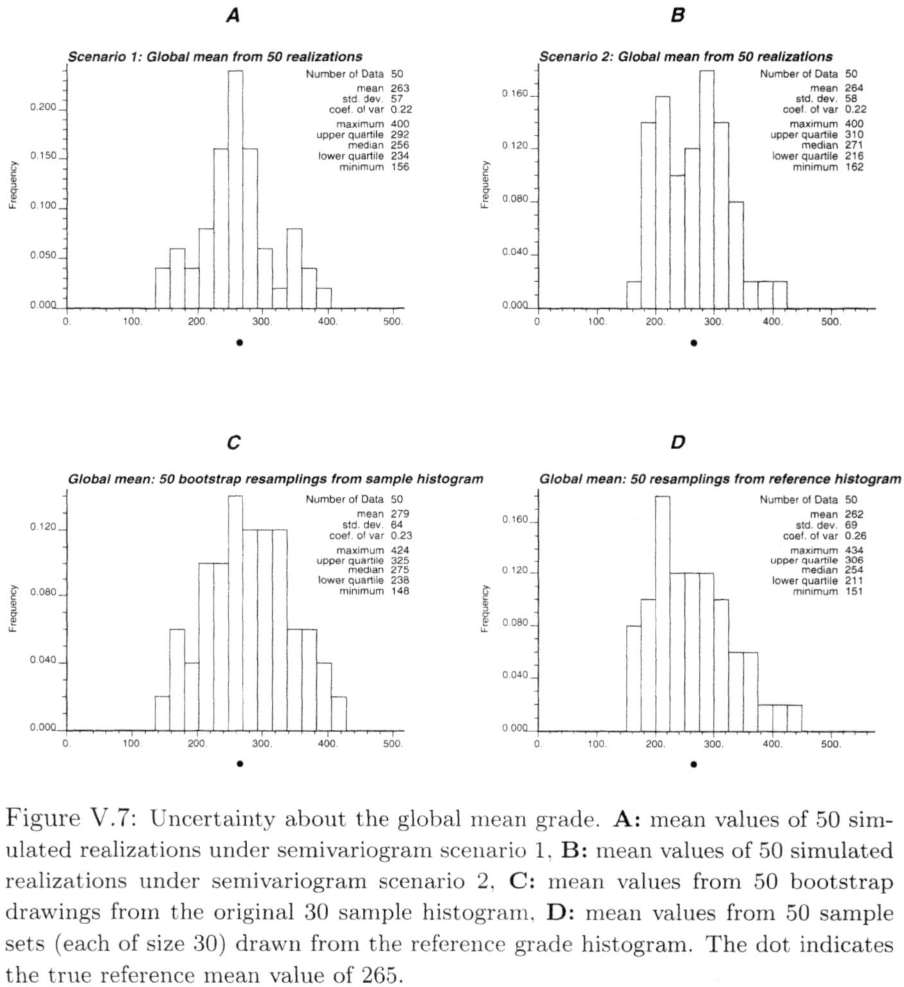

Figure V.7: Uncertainty about the global mean grade. **A:** mean values of 50 simulated realizations under semivariogram scenario 1, **B:** mean values of 50 simulated realizations under semivariogram scenario 2, **C:** mean values from 50 bootstrap drawings from the original 30 sample histogram, **D:** mean values from 50 sample sets (each of size 30) drawn from the reference grade histogram. The dot indicates the true reference mean value of 265.

V.2 Simulation of Perfect Selection

In Section IV.2, perfect selection was performed on the true SMU grades $z_v(\mathbf{u})$, and the same true SMU grades were recovered. In the actual practice of reserves assessment, these SMU grades are not known but could be replaced by simulated SMU grades $z_v^{(l)}(\mathbf{u})$. If there is one single reference recovery indicator map per cutoff (see Fig. IV.7A for cutoff $z = 76$ applied to the true 11×11 SMU grades), there are L corresponding simulated indicator maps if L simulated grade realizations were generated. We will use the following notation for these simulated recovery indicators:

$$i_v^{0,(l)}(\mathbf{u}; z) = \begin{cases} 1 & \text{if } z_v^{(l)}(\mathbf{u}) > z \\ 0 & \text{if not} \end{cases} \tag{V.2}$$

This notation is the simulated counterpart of the notation (I.1) used for the reference indicator $i_v^0(\mathbf{u}; z)$; the superscript 0 denotes perfect selection, and l denotes the lth realization.

Figure V.8 gives the five simulated recovery indicator maps corresponding to the five cutoff grades (IV.3) applied to the simulated 11×11 SMU grades, for realization $l = 6$ with the median mean (that realization is shown in Fig. V.5C). The simulated recovery indicator maps of Figure V.8 should be compared to the corresponding reference indicator maps of Figure IV.7.

The spatial patterns observed in the indicator maps of Figure V.8 are similar to those found in the reference indicator maps. As the cutoff grade increases, however, the SMUs declared as ore are not located accurately over the panel V (compare to the indicator maps of Figs. IV.7D and IV.7E).

For each cutoff grade z and for each SMU $v(\mathbf{u})$, the set of L simulated recovery indicators $\{i_v^{0,(l)}(\mathbf{u}; z),\ l = 1, \ldots, L\}$ provides an estimate of the probability for the SMU $v(\mathbf{u})$ to have true grade $z_v(\mathbf{u}) > z$ (perfect selection). More precisely, that probability is estimated (modeled) by the proportion of simulated recovery indicators valued 1:

$$\text{Prob}\{Z_v(\mathbf{u}) > z|(n)\} = \frac{1}{L} \sum_{l=1}^{L} i_v^{0,(l)}(\mathbf{u}; z) \tag{V.3}$$

Note that these probabilities are conditional to (function of) the $n = 30$ sample data available.

Figure V.9 gives the five maps of the probability values (V.3) corresponding to the five cutoff grade values (IV.3). These probability maps could be used to evaluate where ore is located within panel V; recall that these probabilities relate to perfect selection based on the true SMU grades: They reflect the present uncertainty given the presently available data (here $n = 30$ samples).

Surprisingly enough for such a small sample of $n = 30$ data, these probability maps delineate fairly accurately the extent of the SMUs with grades higher than the corresponding cutoffs (compare to the reference indicator maps of Fig. IV.7).

In addition to the simulated recovery indicators (V.2), one can retrieve from each set of simulated SMU grades $\{z_v^{(l)}(\mathbf{u}), \mathbf{u} \in V\}$ the corresponding simulated recovery functions (again under perfect selection):

- Tonnage recovery within V:

$$t_v^{0,(l)}(z) = \sum_{j=1}^{N} i_v^{0,(l)}(\mathbf{u}_j; z) \ \in \ [0, N] \tag{V.4}$$

- Quantity of metal recovered:

$$q_v^{0,(l)}(z) = \sum_{j=1}^{N} i_v^{0,(l)}(\mathbf{u}_j; z) \cdot z_v^{(l)}(\mathbf{u}_j; z)$$

- Mean ore grade:

$$m_v^{0,(l)}(z) = q_v^{0,(l)}(z)/t_v^{0,(l)}(z)$$

- Profit $p_v^{0,(l)}(z)$: a known function of the simulated tonnage and quantity of metal $t_v^{0,(l)}(z)$ and $q_v^{0,(l)}(z)$.

These formulae are the simulated counterparts of expressions (I.1–I.5).

From the L simulated recovery values at a given cutoff z, say the profit values $\{p_v^{0,(l)}(z), l = 1, \ldots, L\}$, one can retrieve:

- the mean simulated profit:

$$p_v^{0*}(z) = \sum_{l=1}^{L} p_v^{0,(l)}(z)$$

  as an estimate of the true profit value $p_v^0(z)$ under perfect selection [see expression (I.5)], and

- the 95% probability interval defined by the 0.025 and 0.975 quantiles, $a_{.025}(z)$ and $a_{.975}(z)$, of the distribution of the L simulated profit values:

$$\text{Prob}\{p_v^0(z) \in [a_{.025}(z), a_{.975}(z)]\} = 0.95$$

Figure V.10 gives the mean simulated profit and the 95% probability interval along with the true profit function (the latter is copied from Fig. IV.9D). Figure V.10A relates to perfect selection of SMUs of size 11 × 11, and Figure V.10B relates to SMUs of size 22 × 22. Graphs similar to those given in Figure V.10 could be drawn for the three other recovery functions, tonnage, quantity of metal, and mean ore grade, recovered under perfect selection.

It appears that the small ($n = 30$) sample data set suffices to accurately predict the reference profit curve under perfect selection, even for high cutoff grades. Note, however, that the excellent results of Figure V.10 were obtained assuming knowledge of the reference semivariogram model, and using a "good luck" 30 sample data set well-representative of the reference grade distribution (recall the quantile plot of Fig. V.1C).

Also, perfect selection is never achieved in practice; the critical task is to predict actual recovery under imperfect (actual) selection.

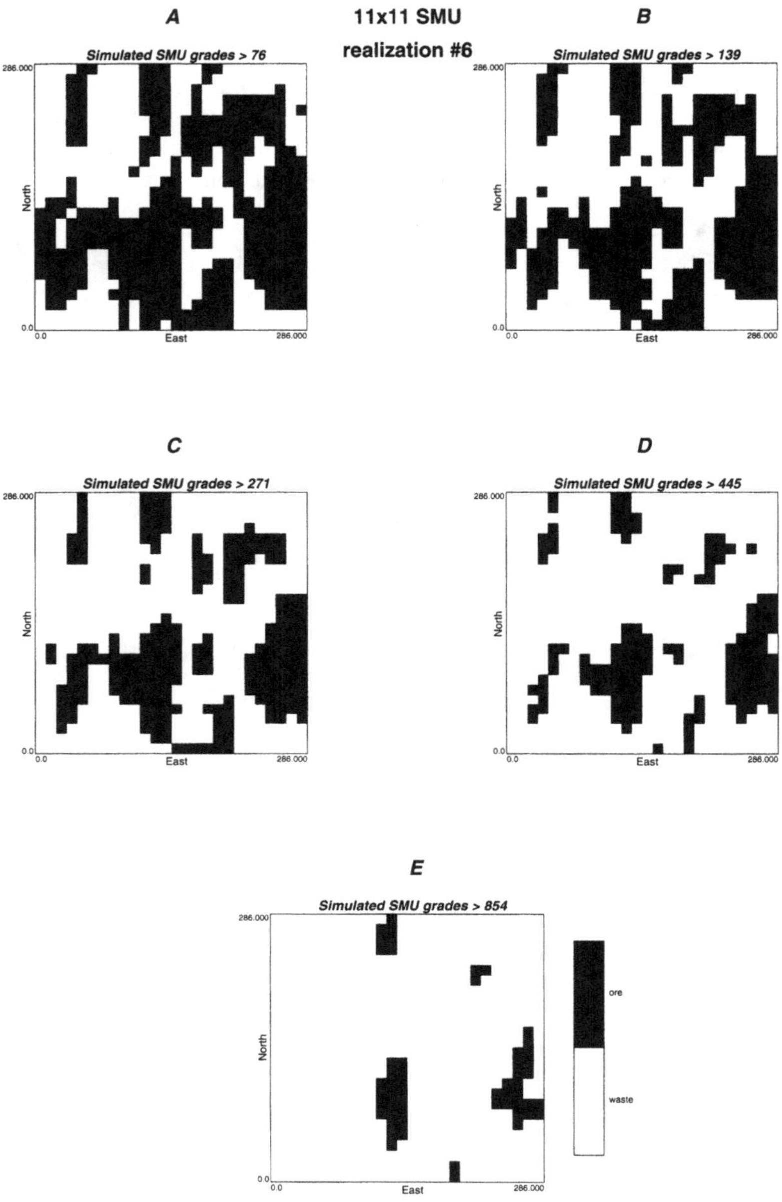

Figure V.8: Maps of simulated recovery indicators under perfect selection for realization #6 (SMU size 11 × 11, reference semivariogram model).

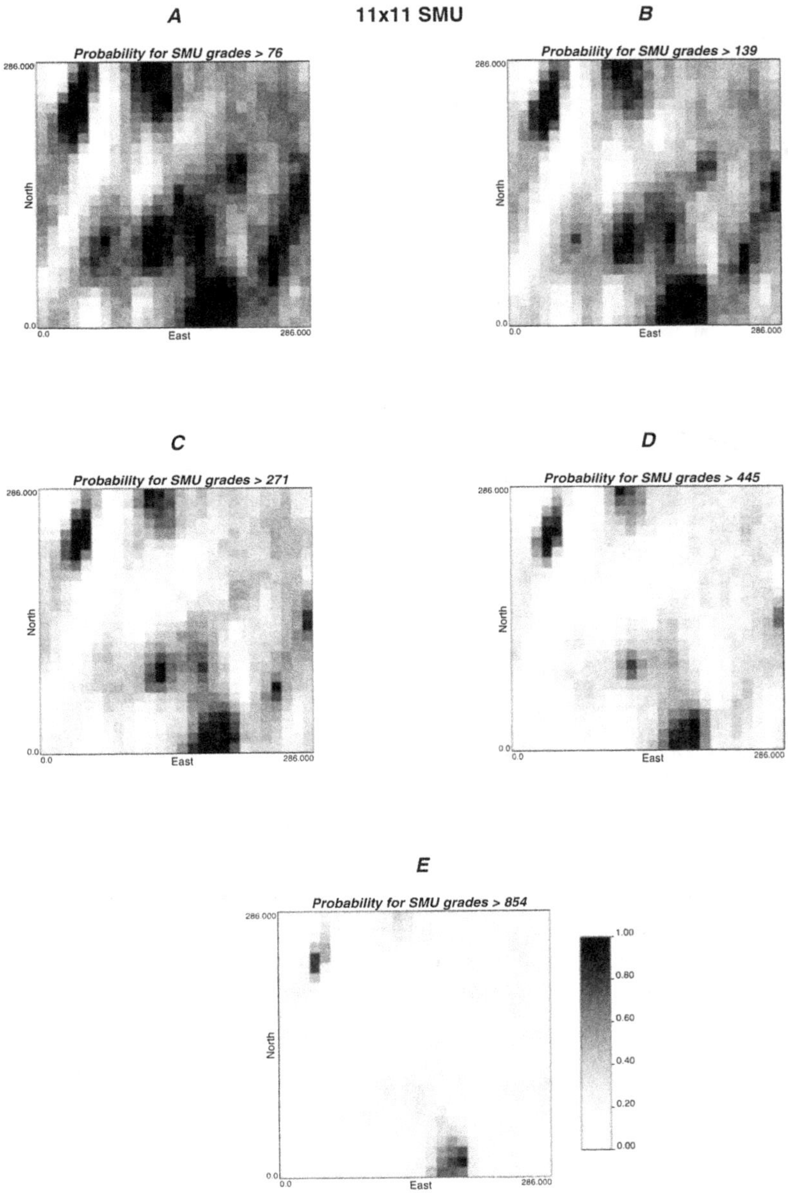

Figure V.9: Probability maps for the true SMU grade to exceed a given cutoff given the present ($n = 30$) sample data, as obtained from 50 simulated SMU grade realizations.

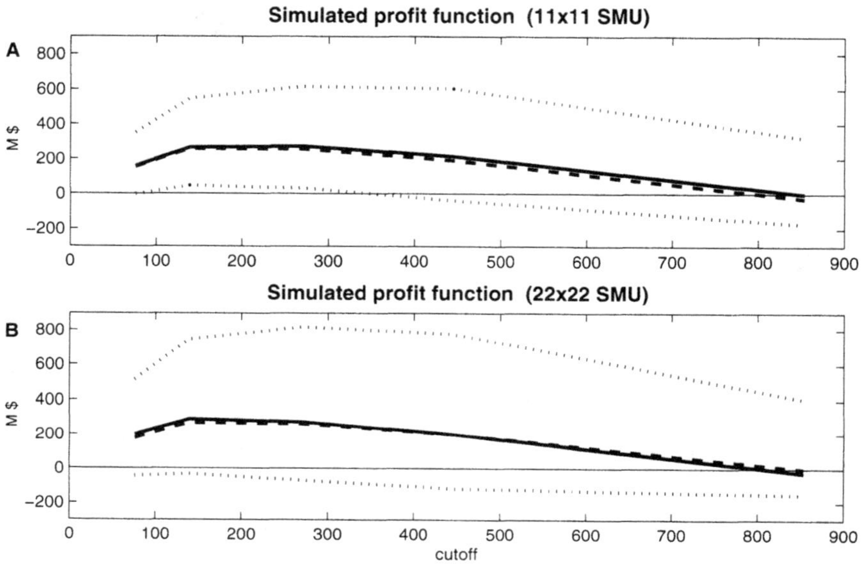

Figure V.10: Simulated profit under perfect selection (Ely data set) for SMU size 11 × 11 (**A**) and 22 × 22 (**B**) under semivariogram model 1 (reference). **Dashed line:** mean simulated profit, **dotted lines:** 95% probability interval from distribution of simulated profit, **solid line:** true profit corresponding to perfect selection.

V.3 Simulation of Actual Selection

Prediction or simulation of recovery assuming perfect selection, as presented in the previous section, is of limited interest: such perfect recovery is never attainable in practice. Instead, SMUs are selected on the basis of their future estimates (Fig. V.11B), themselves derived on future selection data (Fig. V.11A): There will be necessarily some misclassification (recall the discussion in Sections I.3 and II.4).

Simulation of the actual selection process would start with the simulation of the future data (Fig. V.12A) followed by simulation of the future SMU estimation process (Fig. V.12B). A cutoff grade is then applied on these simulated SMU estimates, resulting in simulated recovery indicators (Fig. V.12C). Last, from these simulated recovery indicators the simulated "true" SMU grades are recovered as ore (Fig. V.12D).

There are situations, however, when evaluation of perfect recovery is useful to provide a basis for applying "dilution factors", themselves based on prior mining experience.

Notations (recall)

Simulated "true" SMU grade realizations within any panel V are denoted as $\{z_v^{(l)}(\mathbf{u}), \mathbf{u} \in V\}, l = 1, \ldots, L$. The qualifier "true" differentiates these values from the simulated SMU grade estimates $\{z_v^{**(l)}(\mathbf{u}), \mathbf{u} \in V\}$, $l = 1, \ldots, L$. The superscript $(**)$ serves to recall that these simulated estimates are based on future selection data (of size nn) not available at the present time of the prediction study. These future nn data need to be simulated, and are denoted as $\{z^{(l)}(\mathbf{u}'_\alpha), \alpha = 1, \ldots, nn\}$, $l = 1, \ldots, L$.

The future interpolation/estimation algorithm that is likely to be used at the time of selection is based on the simulated future data yielding the simulated SMU grade estimates $\{z_v^{**(l)}(\mathbf{u}), \mathbf{u} \in V\}$, $l = 1, \ldots, L$, to be used for simulation of the actual selection process. The qualifier actual serves to differentiate that selection process from the "perfect" selection, which would be based on the simulated "true" SMU grades $z_v^{(l)}(\mathbf{u})$.

Simulation can mimic both "true" values and their estimates. Their difference simulates the future misclassification (recall the scattergram of Fig. I.3). It is thus of paramount importance that the notations make that difference clear, even if they are somewhat burdensome.

Future selection data

In this section, the future selection data $\{z(\mathbf{u}'_\alpha), \alpha = 1, \ldots, nn\}$ are assumed to be of the same type, and hence same quality, as the data $\{z(\mathbf{u}_\alpha)$, $\alpha = 1, \ldots, n\}$ presently available. This is different from the simulation of poorer-quality blasthole-type data $b(\mathbf{u}'_\alpha)$ presented in Section IV.3.1 [recall relation (IV.6)]. There is no particular difficulty in attaching to each simulated realization of the true grade $\{z^{(l)}(\mathbf{u}), \mathbf{u} \in V\}$ a realization of the corresponding poorer-quality blasthole field $\{b^{(l)}(\mathbf{u}), \mathbf{u} \in V\}$. We chose not to do so here, to remove this particular source of uncertainty from the final interpretation of the results.

As for the density of the future data, we retained here a 22×22 grid for selection of either 11×11 or 22×22 SMUs. Hence, the smaller SMU size 11×11 may not result in better selectivity because of the coarse 22×22 selection data grid used. That 22×22 grid yields $nn = 169$ selection data values to be simulated; these simulated values are denoted as $\{z^{(l)}(\mathbf{u'}_\alpha), \alpha = 1, \ldots, nn\}$. Recall that the prediction exercise, including the simulation process, is conditioned only to the $n = 30$ present sample data $\{z(\mathbf{u}_\alpha), \alpha = 1, \ldots, n\}$. Figure V.12A displays the simulated future data $\{z^{(l)}(\mathbf{u'}_\alpha), \alpha = 1, \ldots, nn\}$ corresponding to the median realization $l = 6$ shown in Figure V.3C.

Future selection estimates

Retaining only the simulated future data on a 22×22 grid, ordinary block kriging was performed using the GSLIB program `kt3d`; the associated parameter file (for a 11×11 SMU size) is given in Figure V.13. Note that any other interpolation algorithm could have been considered, as long as it is the one expected to be used at the time of selection.

This kriging step results in $L = 50$ sets of simulated SMU estimates $\{z_v^{**(l)}(\mathbf{u}), \mathbf{u} \in V\}$, $l = 1, \ldots, L$. For realization $l = 6$ and for a 11×11 SMU size, these estimates are shown in Figure V.12B. Compare these simulated estimates to the simulated "true" values of Figure V.5C, and note the smoothing effect caused by estimation.

Selection is then performed on the simulated SMU grade estimates $z_v^{**(l)}(\mathbf{u})$, resulting in the simulated recovery indicators:

$$i_v^{**(l)}(\mathbf{u}; z) = \begin{cases} 1 & \text{if } z_v^{**(l)}(\mathbf{u}) > z \\ 0 & \text{if not} \end{cases} \qquad \text{(V.5)}$$

One such simulated recovery indicator map (for realization $l = 6$) is given in Figure V.12C. Last, this simulated indicator map is overlaid over the corresponding simulated "true" SMU grade map (Figure V.5C), resulting in the simulated ore grade map $\{(z_v^{(l)}(\mathbf{u}) | z_v^{**(l)}(\mathbf{u}) > z), \mathbf{u} \in V\}$ shown in Figure V.12D. In this last figure, the SMUs classified as waste are shown as white pixels.

Remark: An interesting alternative to the traditional selection agorithm (V.5) based on the cutoff applied on the SMU estimated grade is that proposed under various similar forms by N.A. Schofield (personal communication, 1996), then Glacken (1997). The probability for the unknown SMU grade $z_v(\mathbf{u})$ to exceed the cutoff grade is derived from the L simulated values $\{z_v^{(l)}(\mathbf{u}), l = 1, \ldots, L\}$. The SMU $v(\mathbf{u})$ is selected as ore if that probability exceeds a location-dependent probability threshold $p_c(\mathbf{u})$, itself determined to minimize the economic loss resulting from potential misclassification.

Simulation of recovery functions

The results displayed in Figure V.12 represent but one simulation ($l = 6$) of the process of *actual* selection and ore recovery displayed in Figure V.11. A set of 49 other such simulations were generated, for a total of $L = 50$.

For each (lth) realization, the following simulated recovery functions are calculated [recall expressions (II.6)]:

- Tonnage recovered: $t_v^{(l)}(z)$
- Quantity of metal recovered: $q_v^{(l)}(z)$
- Mean ore grade: $m_v^{(l)}(z) = q_v^{(l)}(z)/t_v^{(l)}(z)$
- Profit: $p_v^{(l)}(z)$

In the figures, we will give instead these results standardized to the recovery values at zero cutoff (no selection); more precisely:

- the tonnage recovery factor:

$$\%t_v^{(l)}(z) = t_v^{(l)}(z)/t_v(0) \in [0, 1]$$

  where $t_v(0) = N$ is the total number of SMUs in panel V, and

- the quantity of metal recovery factor:

$$\%q_v^{(l)}(z) = q_v^{(l)}(z)/q_v^{(l)}(0) \in [0, 1]$$

  where $q_v^{(l)}(0) = \sum_{j=1}^{N} z_v^{(l)}(\mathbf{u}_j)$.

The mean ore grade $m_v^{(l)}(z)$ and the profit $p_v^{(l)}(z)$ are not standardized, and hence are given in grade unit and $M\$$, respectively.

Results

Figure V.14A gives the histogram of the $L = 50$ simulated values $\%t_v^{(l)}(z)$ of the percent tonnage actually recovered for SMU size 11×11. Figure V.15 gives the results corresponding to SMU size 22×22.

Figures V.16 and V.17 give the histograms of simulated percent quantity of metal values $\%q_v^{(l)}(z)$ actually recovered. Figures V.18 and V.19 give the corresponding histograms of simulated mean ore grade values $m_v^{(l)}(z)$. Last, Figures V.20 and V.21 give the corresponding histograms of simulated actual profit values $p_v^{(l)}(z)$.

The boxplot at the bottom of each histogram depicts the range of variability of the 50 simulated values: the outside whiskers define the 95% probability interval, the inside box defines the 50% probability interval, the vertical solid line in the box gives the median of the 50 simulated values; the black dot gives the true actual recovery value calculated from the reference (true) images of

Figure IV.5, using the same selection constraints (that is, via the procedure described in Fig. V.11).

The true values (black dots) almost always fall within the predicted 50% probability interval, which indicates that the set of simulated recovery values provides an accurate assessment of the related uncertainty.

Figure V.22A gives a plot of the mean simulated actual profit and the 95% probability interval versus the cutoff value z for the 11×11 SMU size. The plot of the true actual profit is also given as a comparison yardstick. Figure V.22B gives the same plots for the 22×22 SMU size. Table V.1 gives the same results in tabulated format. Recall that the reference profit of Table V.1 is not the same as the one given in Table IV.3: The reference profit of Table V.1 is based on better-quality future blasthole data than the reference profit of Table IV.3.

The performance of the proposed simulation approach is quite satisfactory, in that the resulting 50% probability intervals almost always contain the true recovery values (black dots in Figures V.20 and V.21). As expected, the largest probability spread (uncertainty) is generally observed for the largest cutoff $z = 854$. Figure V.22 shows the mean simulated profit curve to match tightly the true profit values: Besides providing a reliable assessment of uncertainty, these simulated values also provide an accurate estimate of the true recovery functions.

11 × 11 SMU			
Cutoff (z)	Reference Profit	Sim. Mean	Sim. 95% PI
76	145	133	[−22, 332]
139	232	219	[12, 501]
271	209	195	[−23, 550]
445	125	114	[−110, 519]
854	−109	−78	[−197, 254]
22 × 22 SMU			
Cutoff (z)	Reference Profit	Sim. Mean	Sim. 95% PI
76	189	179	[24, 381]
139	271	260	[45, 554]
271	238	227	[0, 587]
445	163	146	[−66, 549]
854	−60	−41	[−148, 287]

Table V.1: Reference profit under actual selection based on perfect-quality future blastholes (Ely data set), and statistics of simulated profit with selection performed on estimated SMU grades under semivariogram scenario 1 (reference anisotropic model).

Recall again that the last column of Table IV.3 gave the true profit under actual selection, based on much poorer-quality blasthole data. Comparison of the reference profit values of Tables IV.3 and V.1 provides an assessment of increased profit resulting from better-quality selection data. For example, for SMU size 11 × 11, the true profit obtained at the second cutoff value $z = 139$ is $232M\$$ using the best-quality data; it is only $222M\$$ using the poorer-quality data, corresponding to a 4.3% loss.

For all cutoff values, the 95% probability intervals are extremely large, whereas the 50% probability intervals are more manageable (see Figs. V.20 and V.21). Such large 95% probability intervals are caused by a few simulated realizations that can be considered as outliers, yet are possible realizations given the sample data available and the particular algorithm retained for simulation. One may use additional information to try to cull out such outlier realizations; otherwise, one must accept that at the stringent 95% confidence level there has to be a major risk associated with any mining project. Recall also the large uncertainty about the global mean, as depicted in Figure V.7.

Results under semivariogram scenario 2

All results presented in Figures V.2–V.22 and Table V.1 are based on simulated realizations generated using the semivariogram scenario 1; that is, using the reference normal score semivariogram model given in expression (IV.1). Note that for the block kriging (Fig. V.11B) based on the reference future data (Fig. V.11A), we used the reference z-semivariogram model given in expression (IV.2). In practice, the sparse sample available at the time of prediction may not allow determination of an accurate semivariogram model, not to mention the "true" semivariogram; compare the sample semivariograms of Figure V.1D to the exhaustive reference semivariograms of Figure IV.2B. In this section, we repeated the previous simulation and prediction exercise using the isotropic semivariogram model [expression (V.1)], which was fitted to the actual sample semivariograms of Figure V.1D. Note that model (V.1) is significantly different from the reference model (IV.2). The intent here is to evaluate the effect of poor semivariogram inference on the prediction results. The same original $n = 30$ grade z-values (see Fig. V.1A or V.23B) are used for prediction; the only difference is the semivariogram model adopted.

Figure V.23 gives the grayscale maps of the reference 11×11 SMU grades, and their ordinary kriging estimates from the 30 sample data using semivariogram scenarios 1 and 2, and the corresponding statistics. The differences in the statistics of the estimated values are small, the most notable being the largest maximum estimated value and the larger coefficient of variation when using the isotropic semivariogram model (scenario 2) [compare Figs. V.23D and V.23F].

Simulated realizations

A set of $L = 50$ realizations of point support grade values were generated conditioned to the 30 z-sample data using the isotropic semivariogram model

(V.1). These point support realizations were averaged into simulated realizations of 11×11 SMU grades. Figure V.24 gives the realization with the highest mean value, that with the median mean value, and that with the lowest mean value. This figure should be compared to Figure V.5, which corresponds to the reference semivariogram (scenario 1). The results of Figure V.24 display more rounded (isotropic) structures, as opposed to the more elongated structures of Figure V.5 caused by the anisotropic semivariogram model. Remember, however, that actual selection is not performed on the true SMU grades, but on estimated SMU grades.

Recovery prediction

Similar to what was done under semivariogram scenario 1, we retain from each simulated point support realization $nn = 169$ future data, on a 22×22 grid, for future estimation and selection of 11×11 and 22×22 block size SMUs. Block kriging was performed using these simulated future data, and SMU selection is based on the resulting simulated kriging estimates; however, it is the corresponding simulated true SMU grades that are recovered.

Figures similar to the series of Figures V.14–V.21 were generated, but are not presented to save space. The results are very similar to those obtained under semivariogram scenario 1, indicating little sensitivity of the recovery indicators on the semivariogram model adopted. The reason is that, when enough samples are available (here $nn = 169$ data of core quality at the time of selection), the influence of these data prevails over that of the semivariogram model; also SMU grade estimation and selection is less dependent on the semivariogram model than estimation and selection of point support grades.

11×11 SMU			
Cutoff (z)	Reference Profit	Sim. Mean	Sim. 95% PI
76	145	129	[−13, 334]
139	232	202	[9, 449]
271	209	165	[−41, 523]
445	125	77	[−118, 438]
854	−109	−100	[−198, 123]
22×22 SMU			
Cutoff (z)	Reference Profit	Sim. Mean	Sim. 95% PI
76	189	177	[−14, 383]
139	271	250	[10, 545]
271	238	206	[−45, 560]
445	163	112	[−125, 479]
854	−60	−60	[−196, 169]

Table V.2: Reference profit under actual selection based on perfect-quality future blastholes (Ely data set), and statistics of simulated profit with selection performed on estimated SMU grades under semivariogram scenario 2 (isotropic model).

Figure V.25 and Table V.2 give a summary of profit recovery characteristics under semivariogram scenario 2, similar to the results given in Figure V.22 and Table V.1 for semivariogram scenario 1. The true (reference) actual profit values are the same in both Tables V.1 and V.2, because the same reference anisotropic semivariogram model was used to derive the future SMU grade estimates used for actual selection. It appears that use of the isotropic semivariogram model results in a slightly greater mismatch between the true actual profit and the mean simulated actual profit. In general, however, results are similar, showing the small effect of the isotropic semivariogram model adopted.

Warning: Little effect does not imply no effect. Clearly, if the semivariogram model retained was completely wrong, the prediction could be dramatically different and inaccurate. For example, if a quasi-pure nugget effect semivariogram model was adopted, all SMU grades estimated by kriging would be equal (or very similar), leading to an all-or-nothing selection clearly different from the true actual selection.

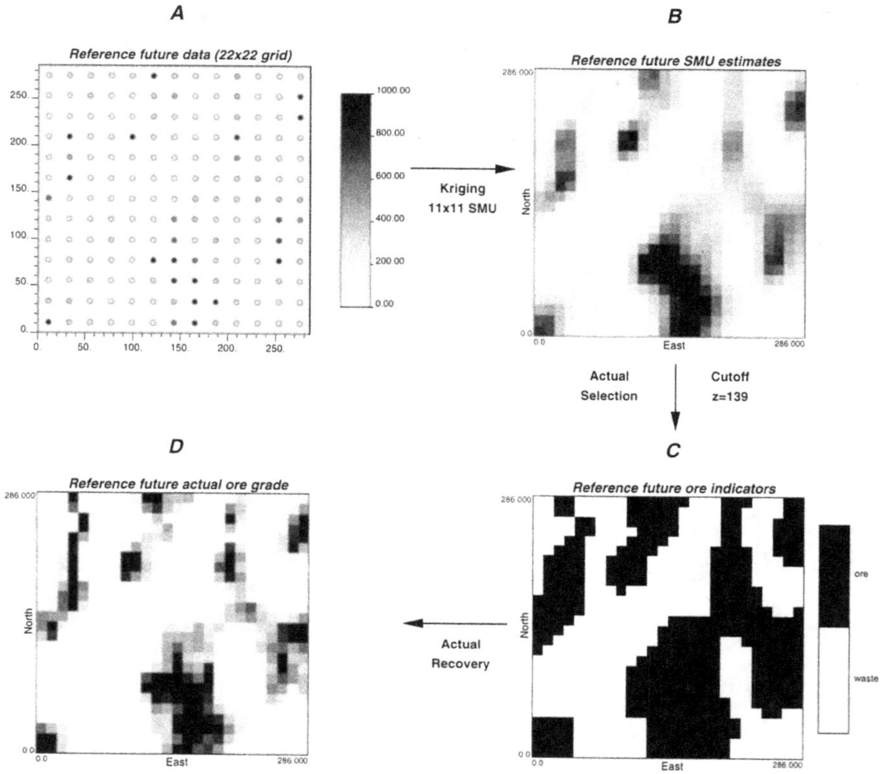

Figure V.11: Actual ore recovery in four steps. **A:** future selection data on a 22×22 grid (taken from the reference data set of Fig. IV.1A), **B:** future 11×11 SMU estimates per kriging, **C:** actual ore indicator map at cutoff $z = 139$, **D:** actual ore grade recovered (true grade values of Fig. IV.5C shown only for the black pixels of **C**; that is, only for ore-declared SMUs).

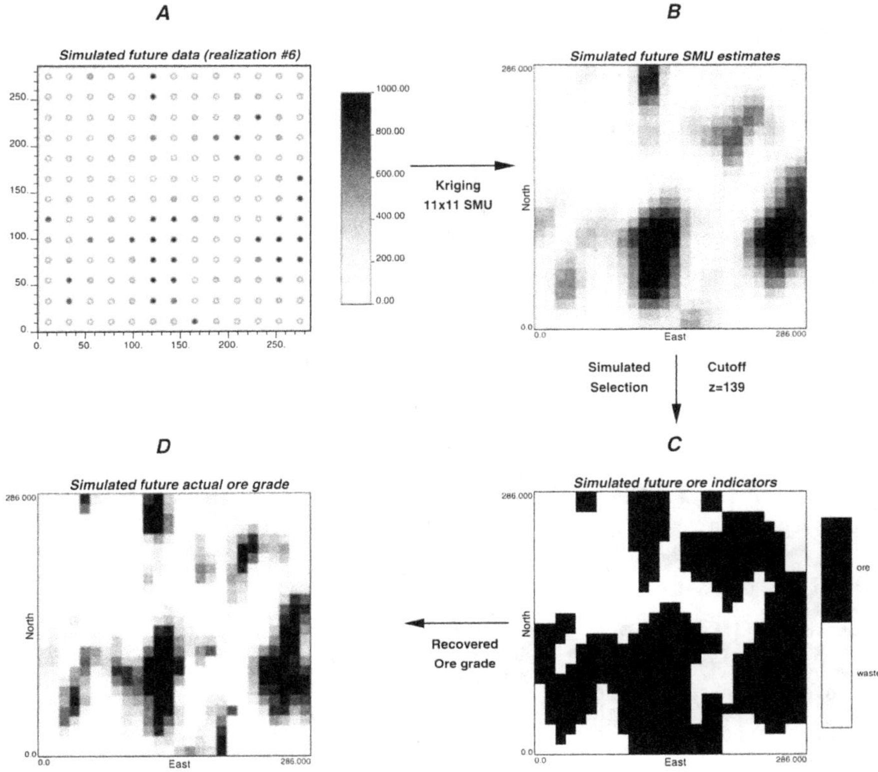

Figure V.12: Simulation of actual ore recovery in four steps. **A:** simulated future selection data (22×22 grid taken from the simulated point support grades of Fig. V.3C), **B:** simulated future 11×11 SMU estimates per kriging, **C:** simulated ore indicator map at cutoff $z = 139$, **D:** simulated actual ore grade recovered (true grade values of Fig. V.5C shown only for the black pixels of **C**; that is, only for ore-declared SMUs).

```
                  Parameters for KT3D
                  *******************

START OF PARAMETERS:
gradecentr22xy.dat                      -file with data
1   2   0    4     0                    -   columns for X, Y, Z, var, sec var
-1.0e21   1.0e21                        -   trimming limits
0                                       -option: 0=grid, 1=cross, 2=jackknife
xvk.dat                                 -file with jackknife data
1   2   0    3     0                    -   columns for X,Y,Z,vr and sec var
1                                       -debugging level: 0,1,2,3
kt3d.dbg                                -file for debugging output
boksmu11.out                            -file for kriged output
26    5.5   11.0                        -nx,xmn,xsiz
26    5.5   11.0                        -ny,ymn,ysiz
1     0.5    1.0                        -nz,zmn,zsiz
4     4      1                          -x,y and z block discretization
2     9                                 -min, max data for kriging
0                                       -max per octant (0-> not used)
33.0  33.0  10.0                        -maximum search radii
 0.0   0.0   0.0                        -angles for search ellipsoid
1      0.000                            -0=SK,1=OK,2=non-st SK,3=exdrift
0 0 0 0 0 0 0 0 0                       -drift: x,y,z,xx,yy,zz,xy,xz,zy
0                                       -0, variable; 1, estimate trend
nodata.dat                              -gridded file with drift/mean
4                                       -  column number in gridded file
2     0.0                               -nst, nugget effect
2     0.65  10.0     0.0    0.0         -it,cc,ang1,ang2,ang3
            70.0    35.0   10.0         -a_hmax, a_hmin, a_vert
2     0.35  10.0     0.0    0.0         -it,cc,ang1,ang2,ang3
            75.0    40.0   10.0         -a_hmax, a_hmin, a_vert
```

Figure V.13: **kt3d** parameter file for kriging the 11×11 SMU grades from simulated future data on a 22×22 grid.

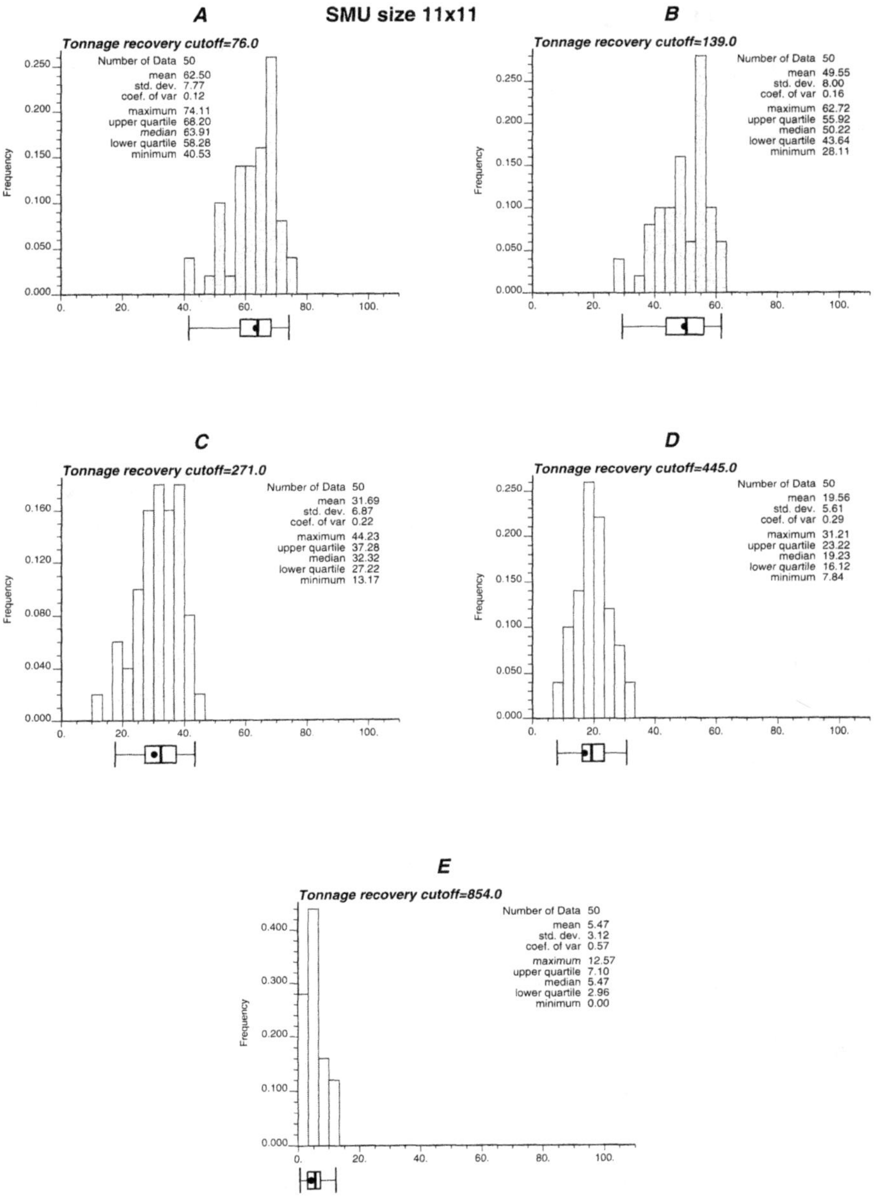

Figure V.14: Simulated percent tonnage recovery (SMU size: 11 × 11, reference semivariogram model).

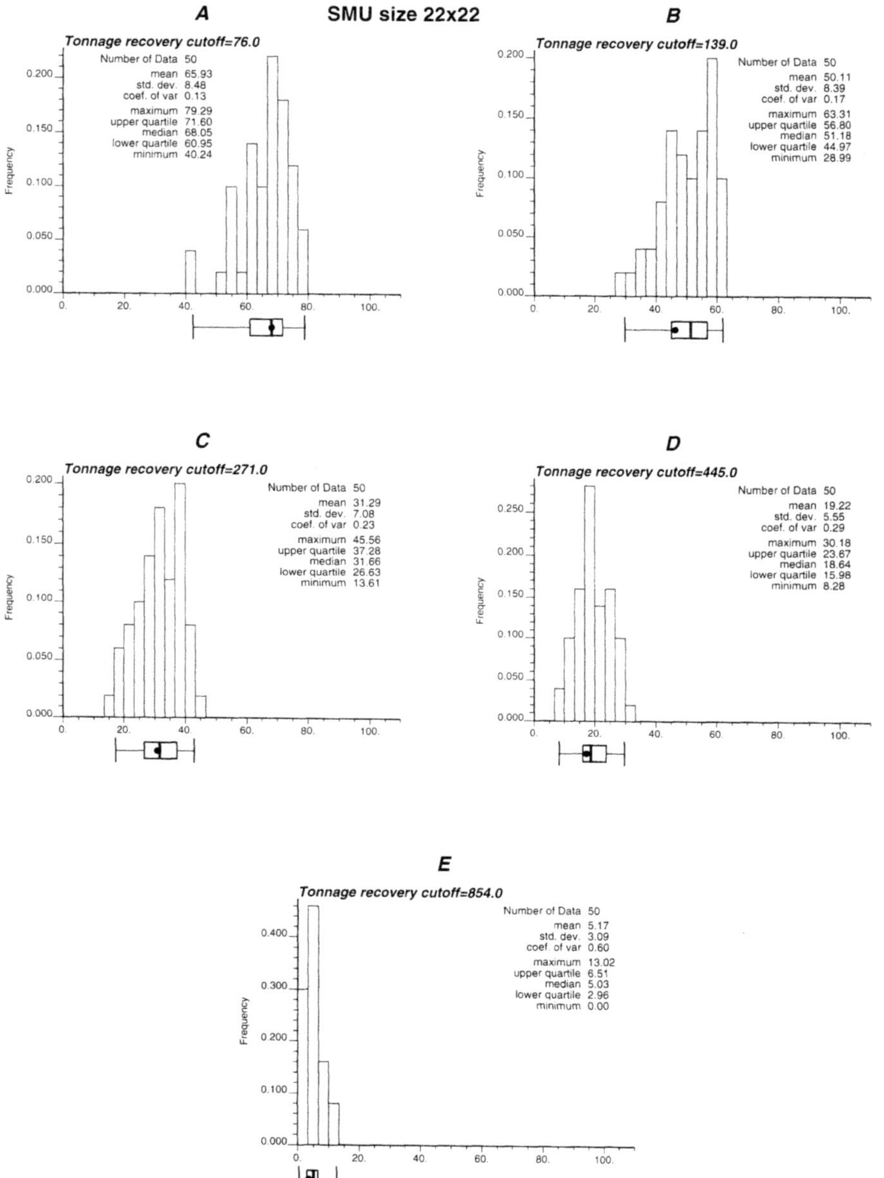

Figure V.15: Simulated percent tonnage recovery (SMU size: 22 × 22, reference semivariogram model).

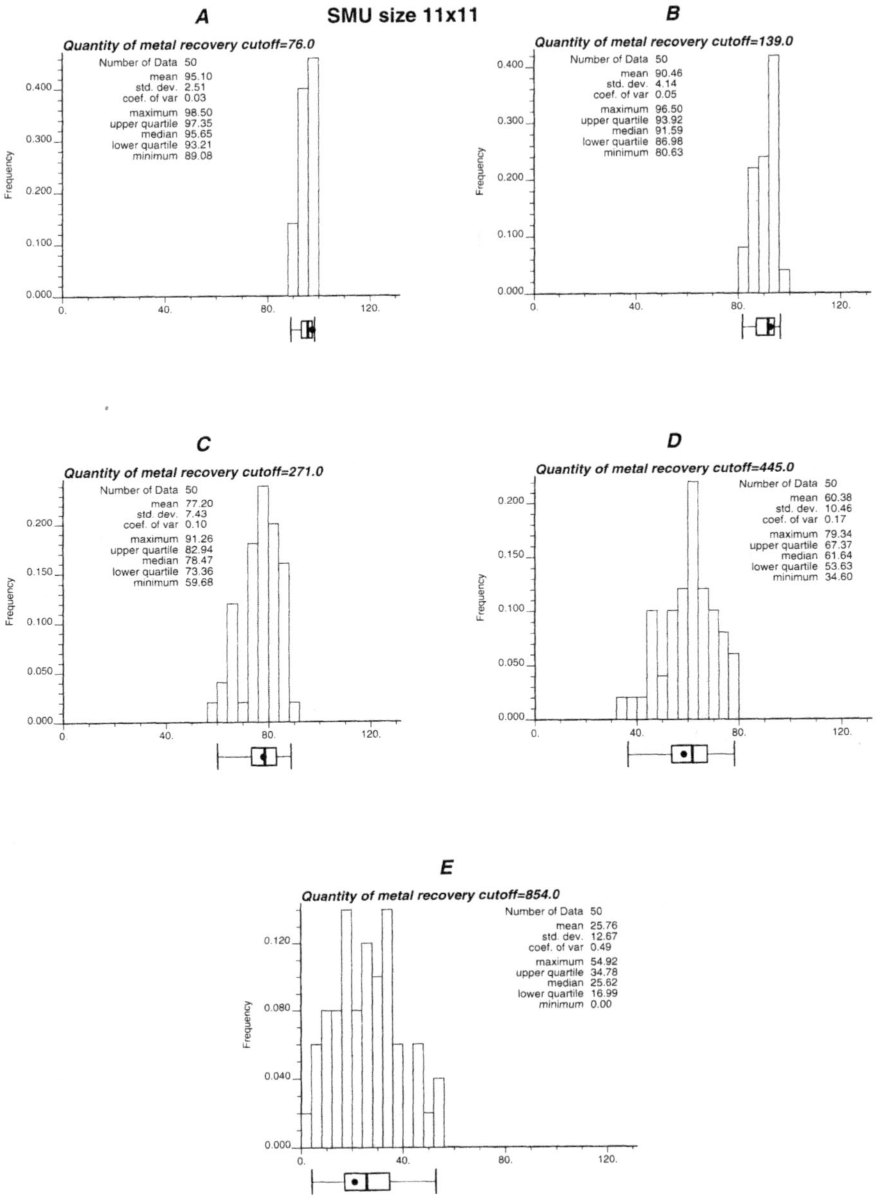

Figure V.16: Simulated percent quantity of metal recovery (SMU size: 11×11, reference semivariogram model).

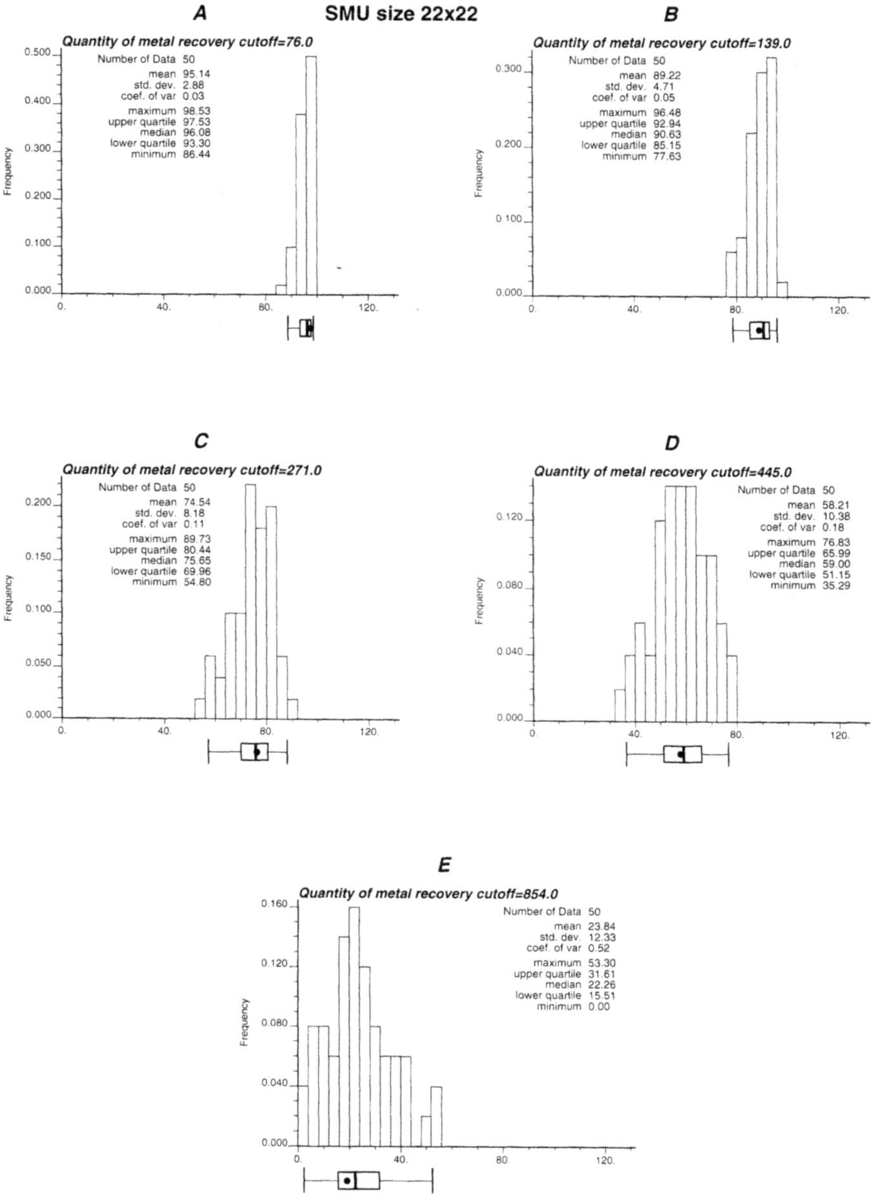

Figure V.17: Simulated percent quantity of metal recovery (SMU size: 22 × 22, reference semivariogram model).

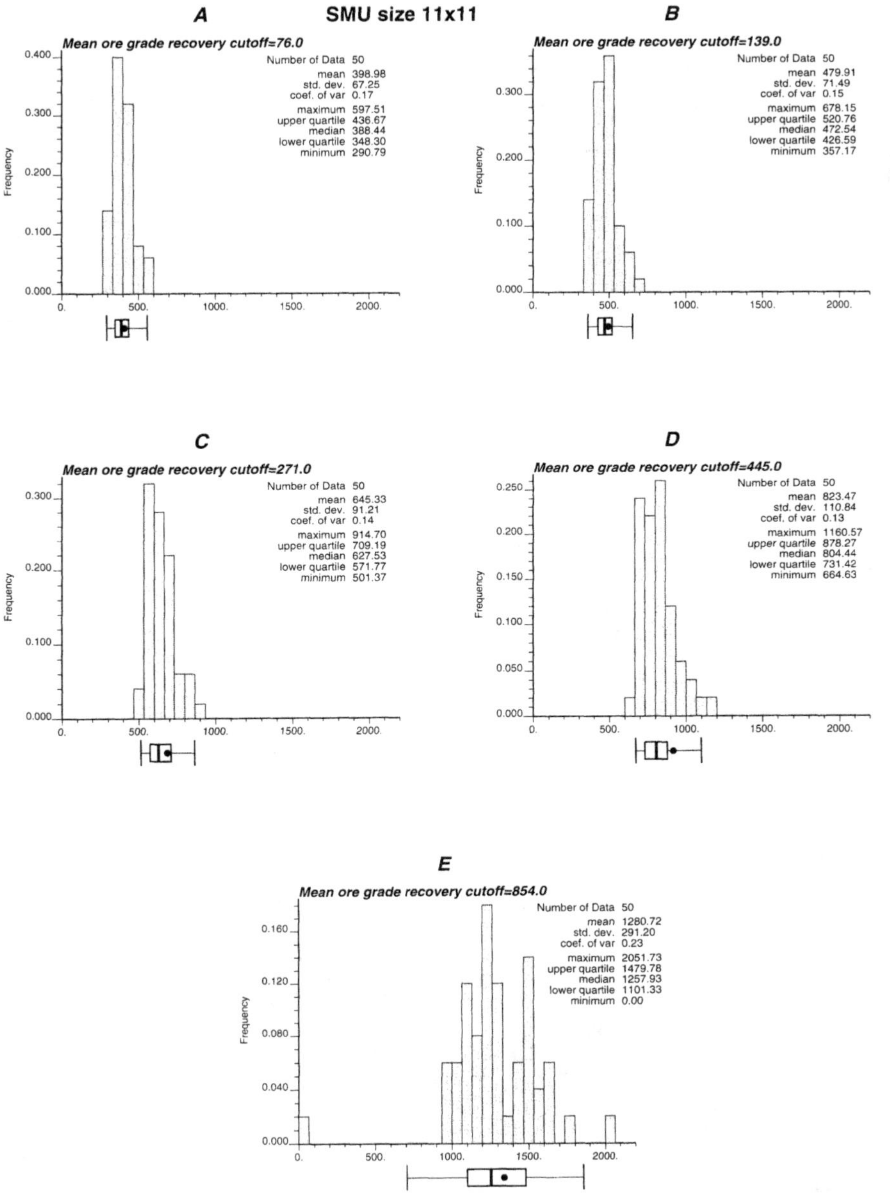

Figure V.18: Simulated mean ore grade recovered (SMU size: 11 × 11, reference semivariogram model).

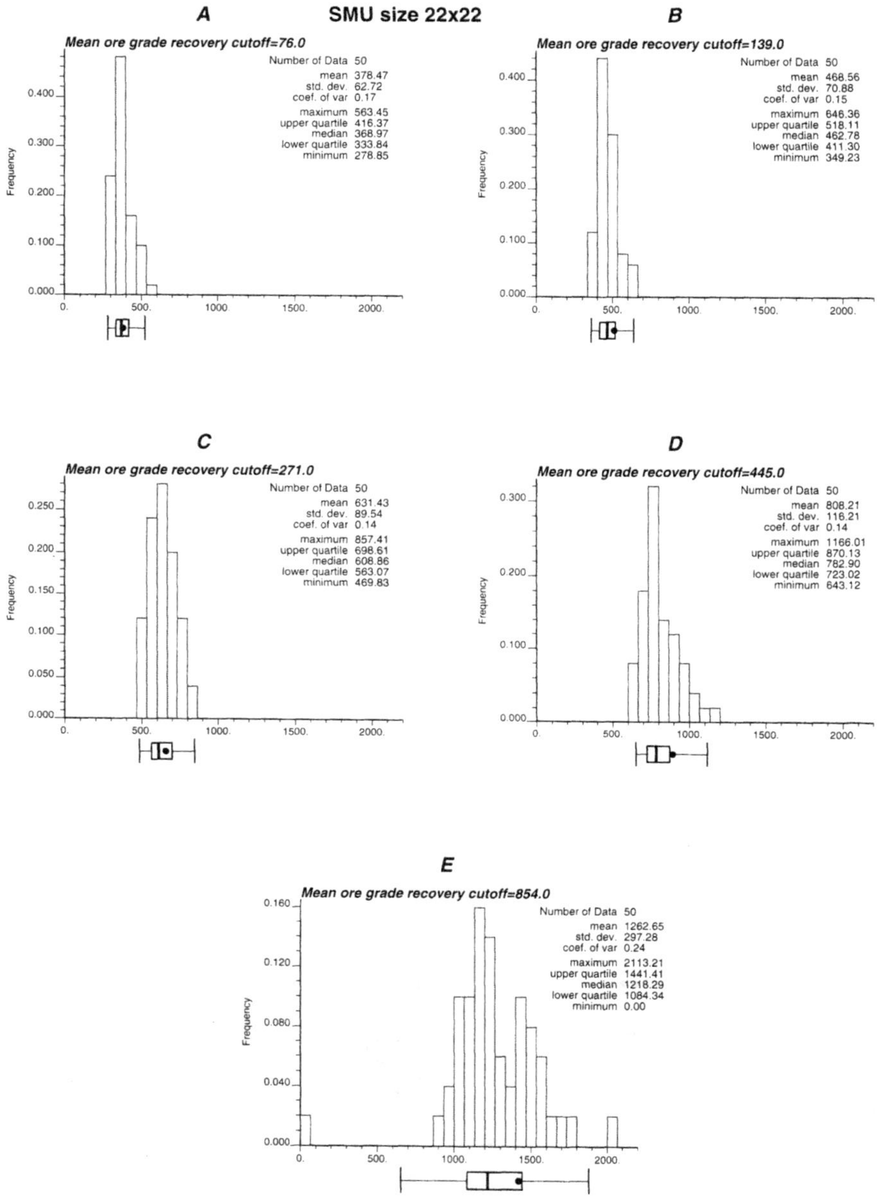

Figure V.19: Simulated mean ore grade recovered (SMU size: 22 × 22, reference semivariogram model).

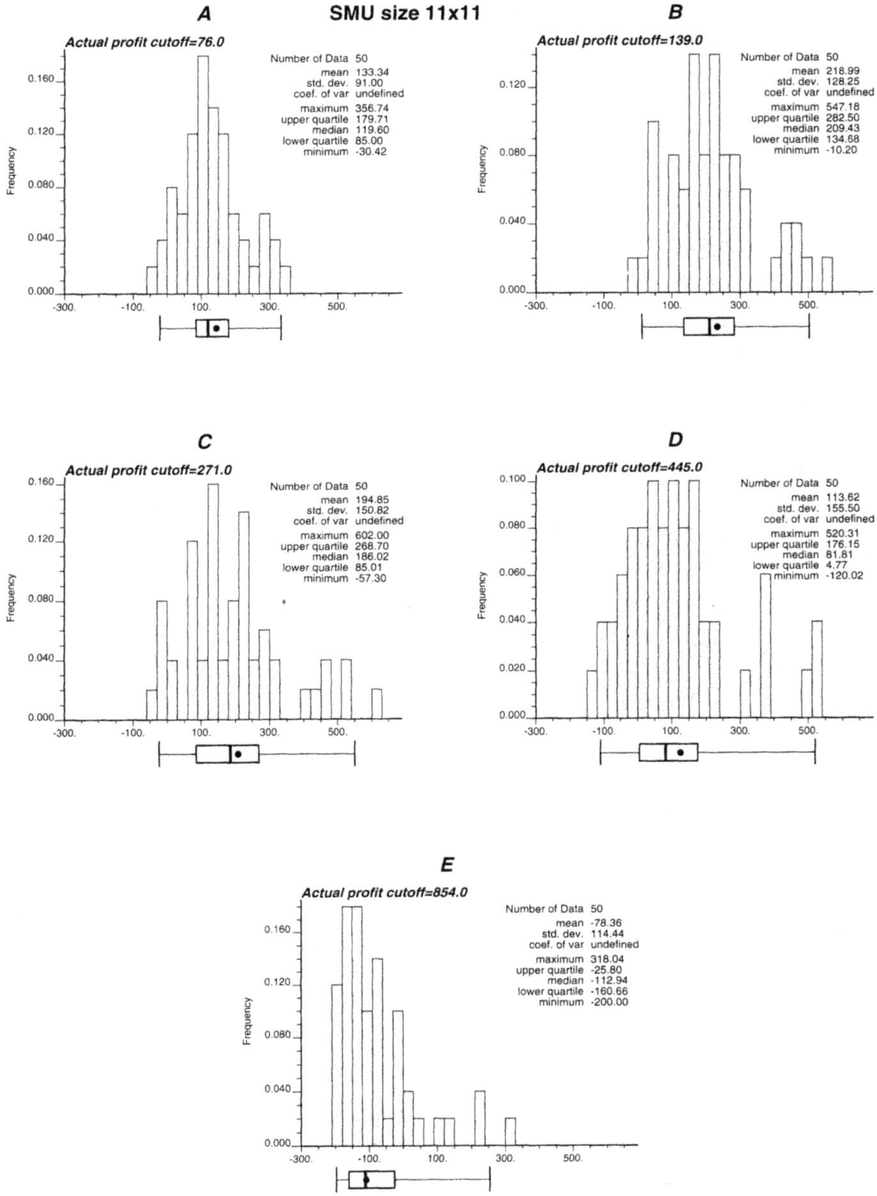

Figure V.20: Simulated actual profit (SMU size: 11×11, reference semivariogram model).

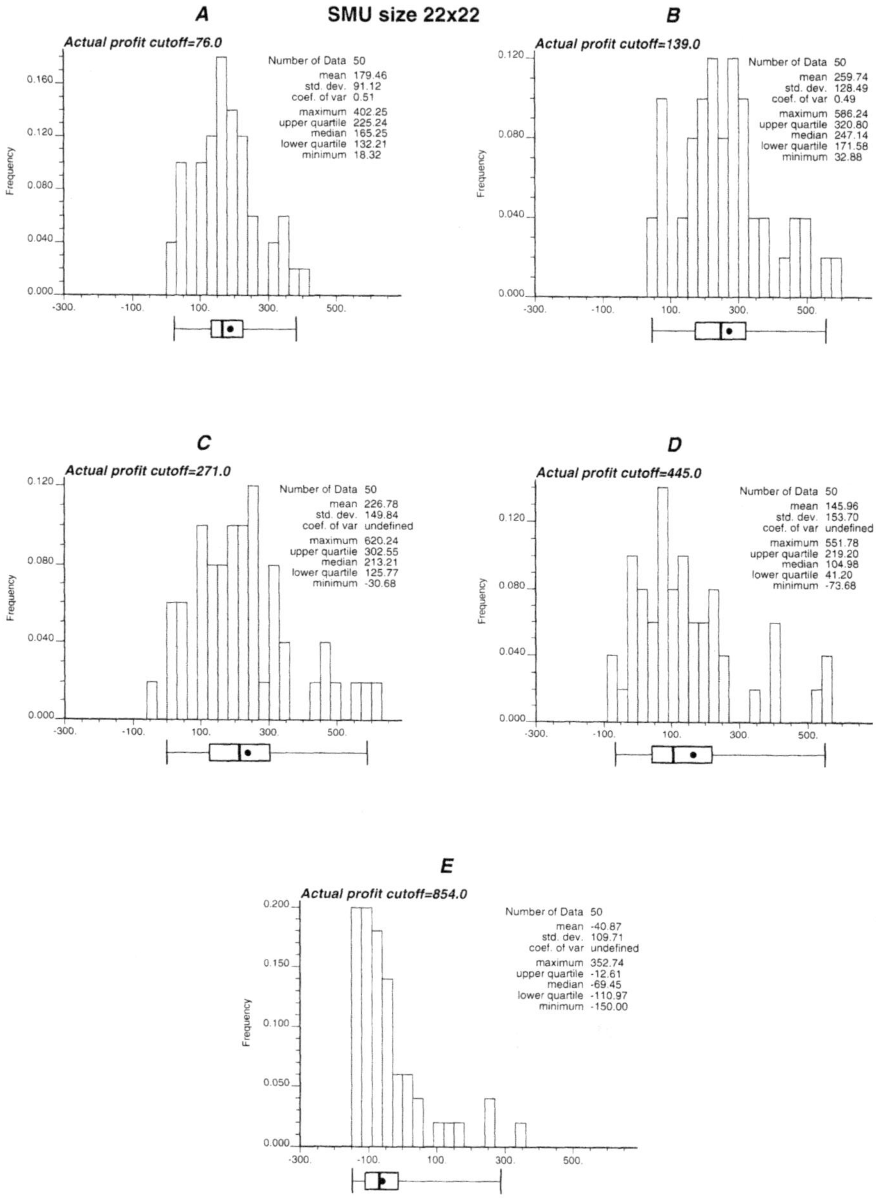

Figure V.21: Simulated actual profit (SMU size: 22×22, reference semivariogram model).

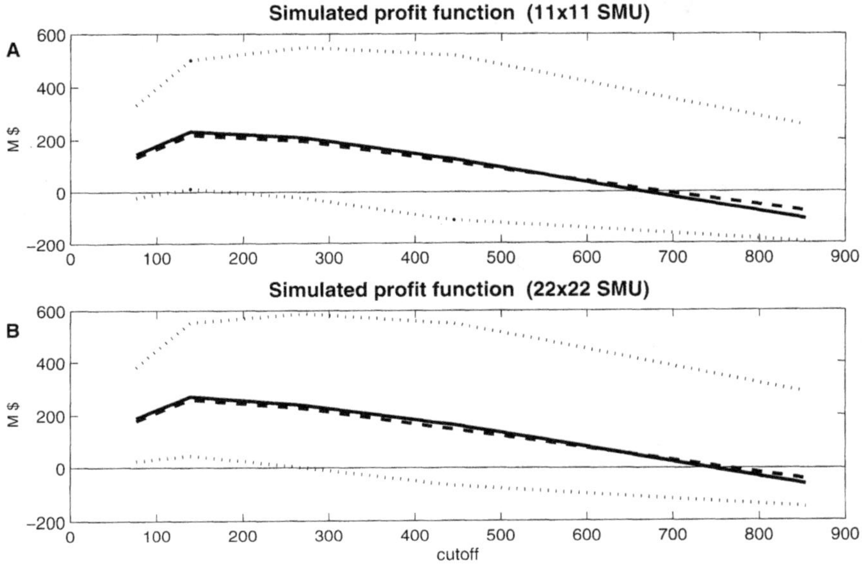

Figure V.22: Simulated actual profit (Ely data set) for SMU size 11×11 (**A**) and 22×22 (**B**) using the reference semivariogram model. **Dashed line:** mean simulated actual profit, **dotted lines:** 95% probability interval from distribution of simulated profit, **solid line:** true profit corresponding to actual selection based on perfect-quality future blastholes.

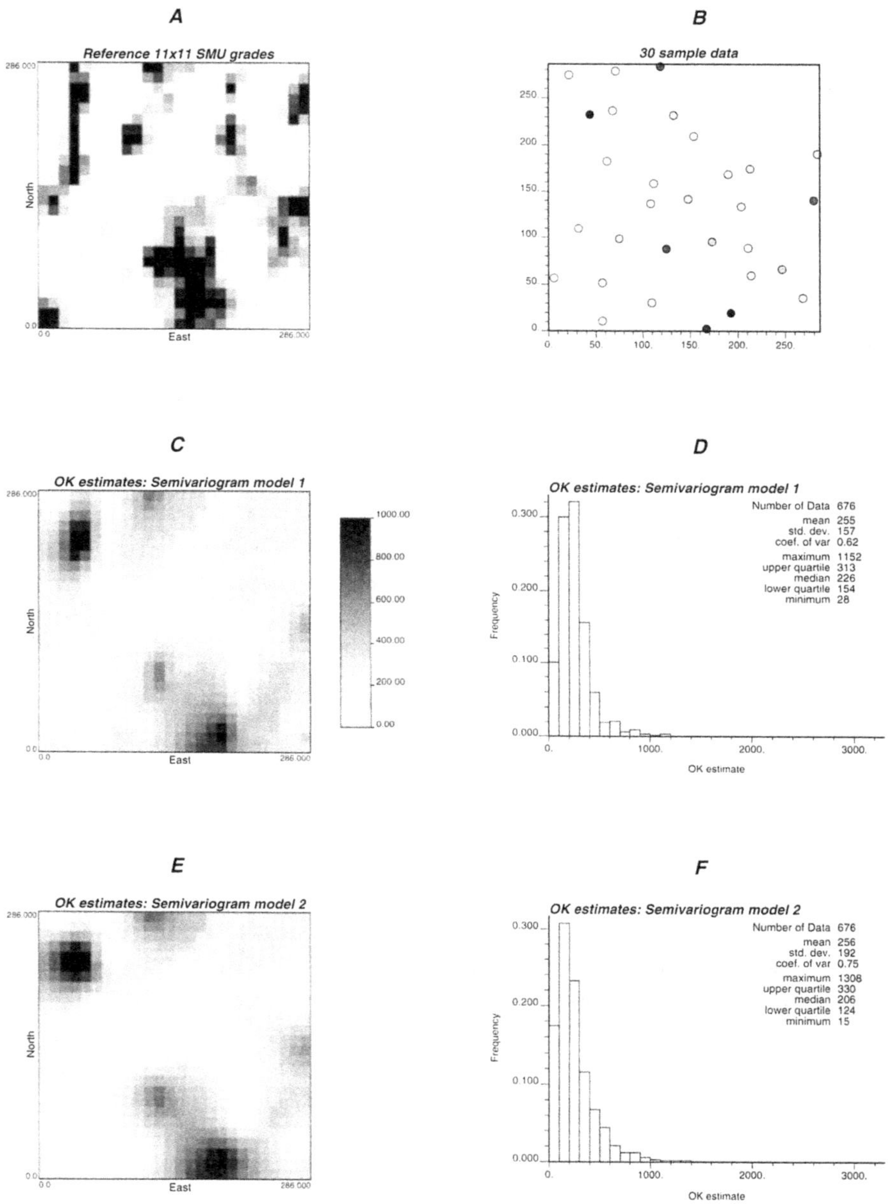

Figure V.23: Reference 11 × 11 SMU grades (**A**) and their ordinary kriging estimates using the 30 sample data (**B**) and, **C**, **D:** semivariogram model 1 (reference-anisotropic), **E**, **F:** semivariogram model 2 (sample-isotropic).

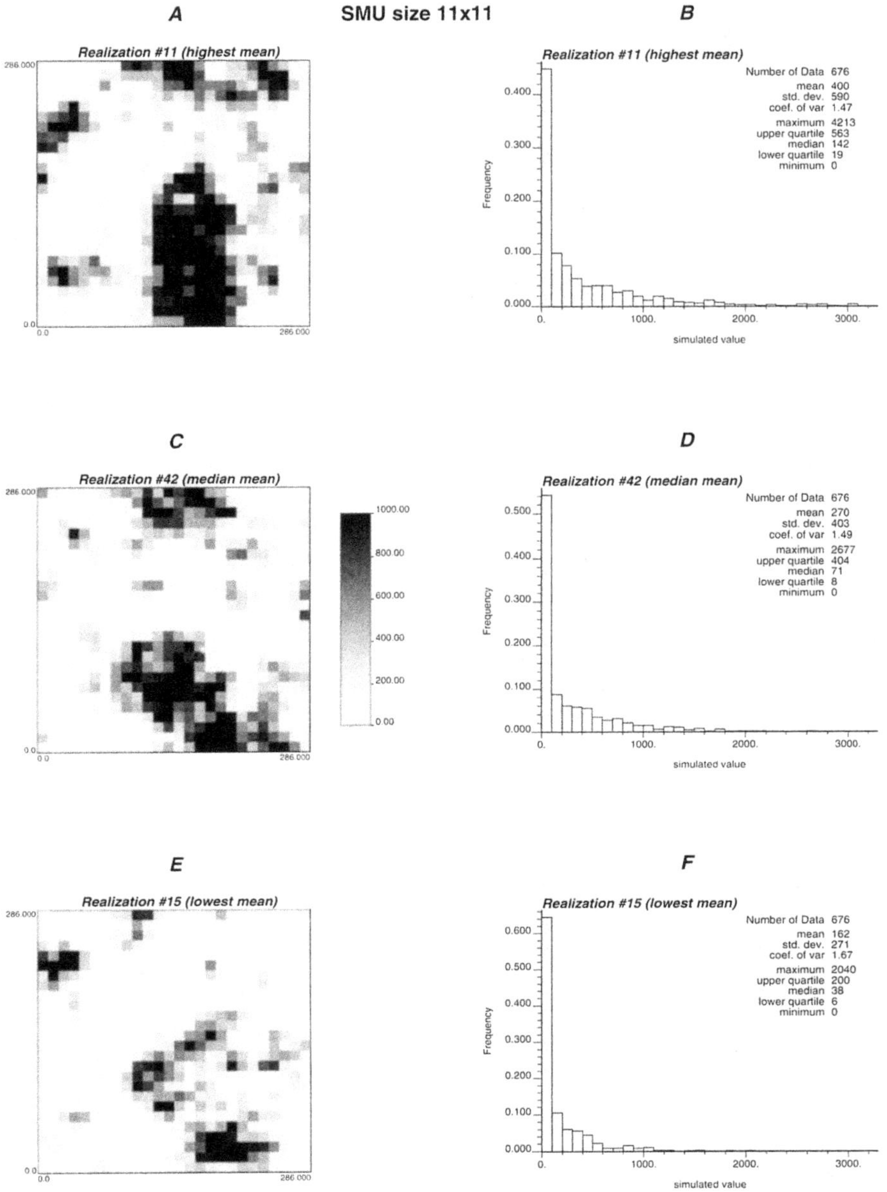

Figure V.24: Three simulated 11×11 SMU grade realizations and their statistics under semivariogram scenario 2 (isotropic model).

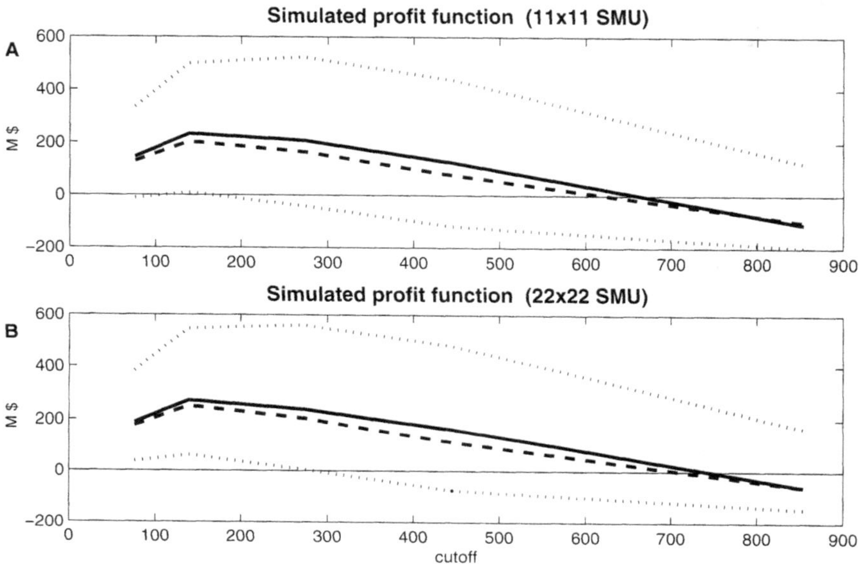

Figure V.25: Simulated actual profit (Ely data set) for SMU size 11×11 (**A**) and 22×22 (**B**) using semivariogram model 2 (isotropic). **Dashed line:** mean simulated actual profit, **dotted lines:** 95% probability interval from distribution of simulated profit, **solid line:** true profit corresponding to actual selection based on perfect-quality future blastholes.

V.4 Application to the Ely2 Data Set

The simulation approach to evaluation of recoverable reserves, as developed in the previous sections, was repeated for the Ely2 data set. This second reference data set is more akin to a two-population gold deposit (recall Figs. IV.19–IV.22).

The information available at the (present) time of the prediction study consists of the original $n = 30$ sample data (see Fig. V.26). The sample locations are the same retained on the Ely deposit (recall Fig. V.1A). Not unexpectedly, from such a sparse sample and the higher variability of the Ely2 reference data set, the sample statistics reflect poorly the reference statistics. Including the zero values, the sample variance is only $(651)^2$ versus the reference variance $(1185)^2$, the quantile-quantile plot of Figure V.26C reflects conditional bias, and the sample normal score semivariograms of Figure V.26D are noisy and fail to reveal any anisotropy. This grim situation is unfortunately not uncommon and portends very uncertain estimates of reserves.

Geological interpretation, for example, from analog deposits, may alleviate the situation by providing indication of the presence of a high-grade second population, its direction of continuity, and the magnitude of grade. Assuming such information is available, the sample indicator semivariograms at the highest economic cutoff grade $z = 854$ were computed along the two directions N10°E and N80°W (see Fig. V.26E). The proportion of sample values greater than this cutoff is $\hat{p} = 0.1$, a value close to the reference proportion $p = 0.09$. The indicator semivariogram model retained includes a 2-to-1 anisotropy ratio reflecting geological interpretation rather than a fit to the sample variograms:

$$\gamma_I(\mathbf{h}) = 0.2 + 0.8\text{Exp}\sqrt{\left(\frac{h_1}{120}\right)^2 + \left(\frac{h_2}{60}\right)^2} \qquad \text{(V.6)}$$

V.4.1 Ely2: Simulation of Point Support Grades

The simulation approach to evaluation of recoverable reserves starts with a simulation of the quasi-point support z-grade field conditional to the $n = 30$ sample z-data. Given knowledge of the presence of two distinct populations, one would start separating them; for example, with a (sequential) indicator simulation using the GSLIB program `sisim` (Deutsch and Journel, 1998, p. 175) and the variogram model (V.6). Figure V.27 gives the `sisim` parameter file used to generate the 50 indicator realizations $\{i^{(l)}(\mathbf{u}), \mathbf{u} \in D\}$. The histogram of the 50 simulated proportions of values $i^{(l)} = 1$ is given in Figure V.28A: The sample proportion $\hat{p} = 0.1$ is correctly bracketed.

Within each of the indicator simulated fields, z-grade simulation is performed separating the two populations:

- The low-grade z-population ($i^{(l)} = 0$) was simulated using sequential Gaussian simulation and the GSLIB program `sgsim`. The normal score variogram used was similar to model (V.6), but with lower nugget effect and longer ranges in both directions (same anisotropy ratio):

$$\gamma_I(\mathbf{h}) = 0.1 + 0.9\text{Exp}\sqrt{\left(\frac{h_1}{140}\right)^2 + \left(\frac{h_2}{70}\right)^2} \qquad (V.7)$$

Again, this model was assumed to be obtained from geological interpretation. The simulation is conditioned to the low-grade sample values and their histogram.

- The high-grade z-population ($i^{(l)} = 1$) was simulated using again program `sgsim`. The normal score variogram used was model (V.6). The simulation is conditioned to the sole high-grade sample values. In early development phases, high-grade sample values are typically too few to allow inference of a reliable histogram. Yet that histogram is needed to evaluate the critical distribution of the high-grade values. In any real practical situation, appropriate hypotheses will be made, possibly capitalizing on prior experience or on available data from deemed-similar deposits. Sensitivity analysis will then have to be performed to appreciate the effect of the z-histogram tail model, (e.g., Parker, 1991; Caers and Maes, 1998). Here we retained the reference tail distribution; that is, the distribution of the reference high-grade population ($z > z_0 = 1000$). This would correspond to an excellent prior guess about the grade tail distribution: Again, geostatistics does not replace good data or good assumptions if the data are not there.

Figure V.28B gives the histogram of the 50 simulated global mean values after recombining low- and high-grade simulated values over each simulated realization. That histogram provides a measure of uncertainty about the global mean within the various hypotheses adopted for the simulation. The 95% probability interval $[208, 632]$ is large and does include both the global sample mean $\hat{m} = 338$ and the reference "true" mean $m = 367$. Figure V.28C gives the histogram of the 50 simulated global coefficient of variation values: $\sigma^{(l)}/m^{(l)}, l = 1, \ldots, 50$. This histogram provides an idea of how uncertain we are about the spatial variability of grades, yet it does not account for the uncertainty related to the hypothesis adopted for the high-grade (tail) distribution.

Figure V.29 gives the grayscale maps and histograms of the three simulated realizations with highest global mean (634), median mean (367), and lowest mean (204). These maps compare poorly to the reference point support grade map of Figure IV.19A, but fortunately reserves are defined on SMU support. Figure V.30 gives the corresponding maps for these realizations averaged over SMUs of size 11×11. When compared to the reference map of Figure IV.22C, it appears again that averaging of uncertainty over block supports is the saving grace of mining predictions.

V.4.2 Ely2: Simulation of Perfect Selection

The simulated point support grade values were averaged into 50 simulated realizations of the 11×11 and 22×22 SMU grades $z_v^{(l)}(\mathbf{u})$. The five cutoff values $z = 76, 139, 271, 445, 854$ were applied successively to each simulated realization to define the simulated recovery functions assuming perfect selection [recall expressions (V.2) and (V.4)].

Figure V.31 gives the mean simulated profit and the 95% probability interval as functions of the cutoff grade. Also given is the true profit function copied from Figure IV.25 and Table IV.4.

Prediction of the true profit is almost as accurate as that achieved for the much easier Ely case (compare to Fig. V.10). Considering the sparse sample ($n = 30$) and the complexity of the Ely2 case, the prediction is rather unexpected: The true profit is close to the mean simulated profit, and is well bracketed by the 95% probability interval obtained from simulation. Recall, though, that we had access to an excellent model for the long tail distribution of high grades.

V.4.3 Ely2: Simulation of Actual Selection

As was done for the Ely data set (recall Section V.3), simulation of the actual selection process starts with the simulation of the future blasthole data followed by simulation of the future estimation process of the SMU grades. The cutoff grade is then applied to these simulated SMU grades, resulting in simulated recovery indicators and functions.

Future blasthole data

Blasthole data for the Ely2 deposit were generated in Section IV.4.3. These are "true" blasthole data as would be available at the time of selection and mining. At the (present) time of prediction, these future actual b-data are not available; instead, they have to be simulated starting from the sole information available; that is, the $n = 30$ original sample values and the simulated realizations of point support grades as displayed in Figure V.29.

For the future blasthole simulation, we should retain a generation process similar to that used for the "true" blasthole grades, thus assuming that we guessed correctly the quality of these future blasthole data. We therefore considered in Table V.3 parameter values $\lambda(\mathbf{u})$ and $\nu(\mathbf{u})$ that are not significantly different from those used for the generation of the "true" blasthole field as listed in Table IV.5. This amounts to assuming that the bias and error variance of the future b-data could be evaluated using some reconciliation data. As in the case of the "true" blasthole field, the local error mean, $m_R(\mathbf{u})$, and the local error variance, $s_R^2(\mathbf{u})$, depend on the distribution of the local window mean $m_W^{(l)}(\mathbf{u})$ and standard deviation $s_W^{(l)}(\mathbf{u})$ of each simulated point support grade realization $\{z^{(l)}(\mathbf{u}), \mathbf{u} \in D\}$. Note that to each different realization $\{z^{(l)}(\mathbf{u}), \mathbf{u} \in D\}$ of the point support grade, corresponds a differ-

$m_W^{(l)}(\mathbf{u})$	$\lambda(\mathbf{u})$	$s_W^{(l)}(\mathbf{u})$	$\nu(\mathbf{u})$
$< [F_{mW}^{-1}(0.2)]^{(l)}$	1.5	$< [F_{sW}^{-1}(0.2)]^{(l)}$	1.0
$[F_{mW}^{-1}(0.2)]^{(l)}$–$[F_{mW}^{-1}(0.4)]^{(l)}$	1.0	$[F_{sW}^{-1}(0.2)]^{(l)}$–$[F_{sW}^{-1}(0.4)]^{(l)}$	0.5
$[F_{mW}^{-1}(0.4)]^{(l)}$–$[F_{mW}^{-1}(0.6)]^{(l)}$	0.0	$[F_{sW}^{-1}(0.4)]^{(l)}$–$[F_{sW}^{-1}(0.6)]^{(l)}$	0.5
$[F_{mW}^{-1}(0.6)]^{(l)}$–$[F_{mW}^{-1}(0.8)]^{(l)}$	−0.5	$[F_{sW}^{-1}(0.6)]^{(l)}$–$[F_{sW}^{-1}(0.8)]^{(l)}$	0.5
$> [F_{mW}^{-1}(0.8)]^{(l)}$	−1.5	$> [F_{sW}^{-1}(0.8)]^{(l)}$	1.0

Table V.3: Ely2 parameters $\lambda(\mathbf{u})$ and $\nu(\mathbf{u})$ used for generating the future blasthole data. These parameter definitions are similar to the ones used for the generation of the reference "true" blasthole data listed in Table IV.5. $F_{mW}^{(l)}(\cdot)$ and $F_{sW}^{(l)}(\cdot)$ are the realization-specific cdfs of local window means and standard deviations.

ent realization $\{b^{(l)}(\mathbf{u}), \mathbf{u} \in D\}$ of the future blasthole grade field. Otherwise, the simulation of the future blasthole data follows the same algorithm presented in Section IV.3.1, expressions (IV.6) to (IV.8). These future b-data need only be simulated over the 22×22 grid expected to be available at the time of actual selection.

Future selection estimates

Retaining only the simulated future b-data on the 22×22 grid, indicator kriging was performed to estimate each SMU grade. Again, we should not consider the same kriging process used in Section IV.4.3 to generate the reference future selection estimates $z^{**}(\mathbf{u})$. Instead, we should simulate a process that can only mimic the actual future selection estimation process.

The simulated indicator kriging process is an ordinary block indicator kriging, which yields a simulated estimate $p^{**(l)}(\mathbf{u})$ of the proportion of point grade values below the highest economical cutoff grade $z = 854$, for each SMU $v(\mathbf{u})$ and each realization number l. The indicator variogram used is model (V.6), the only one available at the (present) time of reserves assessment. A large neighborhood of $7 \times 7 = 49$ future b-data locations was considered.

Next, the simulated mean SMU grade $z_2^{**(l)}(\mathbf{u})$ above the cutoff $z = 854$ was estimated by block kriging, using only the high-grade simulated future b-data. Similarly, the simulated mean SMU grade $z_1^{**(l)}(\mathbf{u})$ below the cutoff $z = 854$ was estimated by block kriging, using only the low-grade simulated future b-data. Last, the three previous simulated estimates were combined into a simulated SMU grade future estimate using a formula similar to (IV.17):

$$z_v^{**(l)}(\mathbf{u}) = z_1^{**(l)}(\mathbf{u})p^{**(l)}(\mathbf{u}) + z_2^{**(l)}(\mathbf{u})\left[1 - p^{**(l)}(\mathbf{u})\right]$$

Simulation of the future selection is then performed on the simulated SMU grade estimates $z_v^{**(l)}(\mathbf{u})$, resulting in simulated recovery indicators and functions.

Results

Figure V.32 gives the histograms of the $L = 50$ simulated percent tonnage recovery for the SMU size 11×11. Figures V.33–V.35 give the histograms of simulated percent quantity of metal, mean ore grade, and actual profit values. The results for the 22×22 SMU size are not given to save space. The very large 95% probability intervals do bracket the corresponding "true" actual recovery values (black dots).

Figure V.36 gives the plot of the mean simulated actual profit and the 95% probability intervals versus the cutoff value z, for both SMU sizes. The plot of the "true" reference profit, taken from Table IV.6, is also given as a comparison yardstick. Table V.4 gives the same results in tabulated format. Recall that the future selection process simulated is not identical to the actual future selection; such inaccuracy in modeling the future selection process is to be expected in any real practical situation.

The performance of the proposed simulation approach in predicting future recoveries under the process of *actual* selection (as opposed to perfect selection) is deemed satisfactory, similar to the performance obtained for the original Ely data set (recall Figs. V.22 and V.25). The reference actual profit is underestimated by the mean simulated profit, especially for the highest cutoff grades, but it is well bracketed by the 95% probability interval. This underestimation could be attributed to the slight mismatch between the parameters used to simulate the future blasthole samples (see Table V.3) and those used to generate the reference future blasthole field (see Table IV.5).

11 × 11 SMU			
Cutoff (z)	Reference Profit	Sim. Mean	Sim. 95% PI
76	305	275	[17, 678]
139	440	395	[34, 968]
271	396	385	[−24, 1095]
445	372	328	[−103, 1093]
854	154	98	[−230, 803]
22 × 22 SMU			
Cutoff (z)	Reference Profit	Sim. Mean	Sim. 95% PI
76	354	327	[67, 740]
139	481	444	[84, 1030]
271	428	421	[28, 1134]
445	366	363	[−65, 1111]
854	194	137	[−156, 816]

Table V.4: Ely2: Reference profit under actual selection and mean simulated profit with selection performed based on simulated future blasthole data.

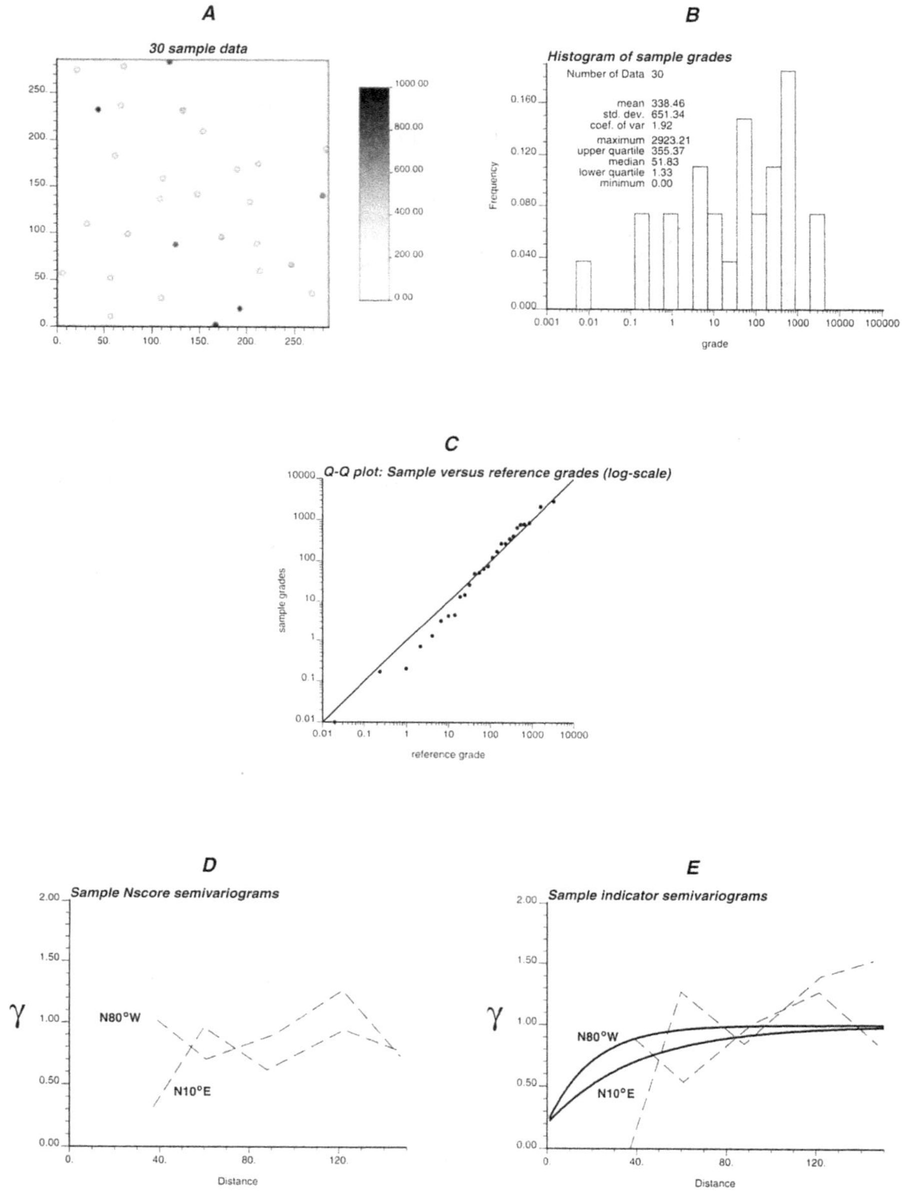

Figure V.26: Ely2 sample data and statistics. **A**, **B**: sample locations and histogram, **C**: quantile plot of sample versus reference distributions, **D**: semivariogram of sample normal score grades, **E**: sample indicator semivariogram at cutoff $z = 854$, standardized to unit sill.

```
                    Parameters for SISIM
                    ********************

START OF PARAMETERS:
0                             -1=continuous(cdf), 0=categorical(pdf)
2                             -number thresholds/categories
1.00  0.00                    -   thresholds / categories
0.10  0.90                    -   global cdf / pdf
sample.dat                    -file with data
1   2   0   4                 -   columns for X,Y,Z, and variable
nodata.dat                    -file with soft indicator input
0   0   0   0 0 0 0 0         -   columns for X,Y,Z, and indicators
0                             -   Markov-Bayes simulation (0=no,1=yes)
0.00  0.00  0.00  0.00  0.00  -      calibration B(z) values
-1.0e21     1.0e21            -trimming limits
0.0   30.0                    -minimum and maximum data value
1       0.0                   -   lower tail option and parameter
1       1.0                   -   middle     option and parameter
1       1.0                   -   upper tail option and parameter
cluster.dat                   -   file with tabulated values
3   0                         -      columns for variable, weight
0                             -debugging level: 0,1,2,3
sisim.dbg                     -file for debugging output
sisim50.out                   -file for simulation output
50                            -number of realizations
286    0.5     1.0            -nx,xmn,xsiz
286    0.5     1.0            -ny,ymn,ysiz
1    1.0    10.0              -nz,zmn,zsiz
32039255                      -random number seed
24                            -maximum original data  for each kriging
12                            -maximum previous nodes for each kriging
1                             -maximum soft indicator nodes for kriging
1                             -assign data to nodes? (0=no,1=yes)
1      3                      -multiple grid search? (0=no,1=yes),num
0                             -maximum per octant    (0=not used)
120.0   60.0   20.0           -maximum search radii
 10.0    0.0    0.0           -angles for search ellipsoid
0     2.5                     -0=full IK, 1=median approx. (cutoff)
0                             -0=SK, 1=OK
1     0.20                    -ONE: nst, nugget effect
2     0.80 10.0    0.0    0.0 -it,cc,ang1,ang2,ang3
         120.0    60.0  10.0  -a_hmax, a_hmin, a_vert
```

Figure V.27: **sisim** parameter file for generating 50 realizations of the indicator field separating low- and high-grade values, conditional on the Ely2 sample grades.

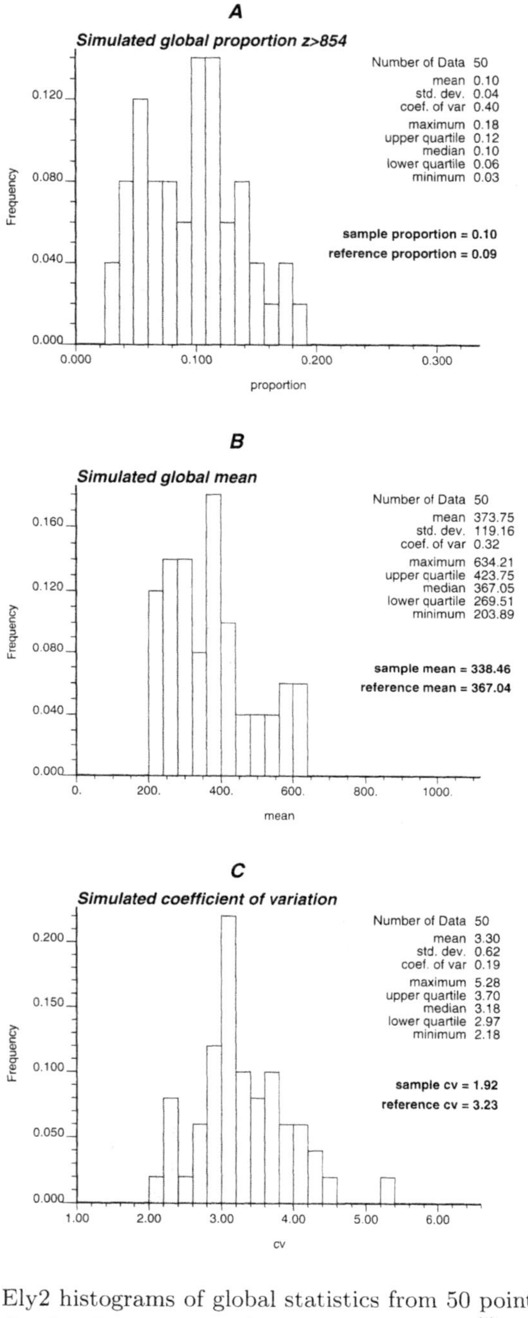

Figure V.28: Ely2 histograms of global statistics from 50 point support grade realizations. **A:** simulated proportion of high-grade values ($z^{(l)} > 854$), **B:** simulated global mean, **C:** simulated global coefficient of variation.

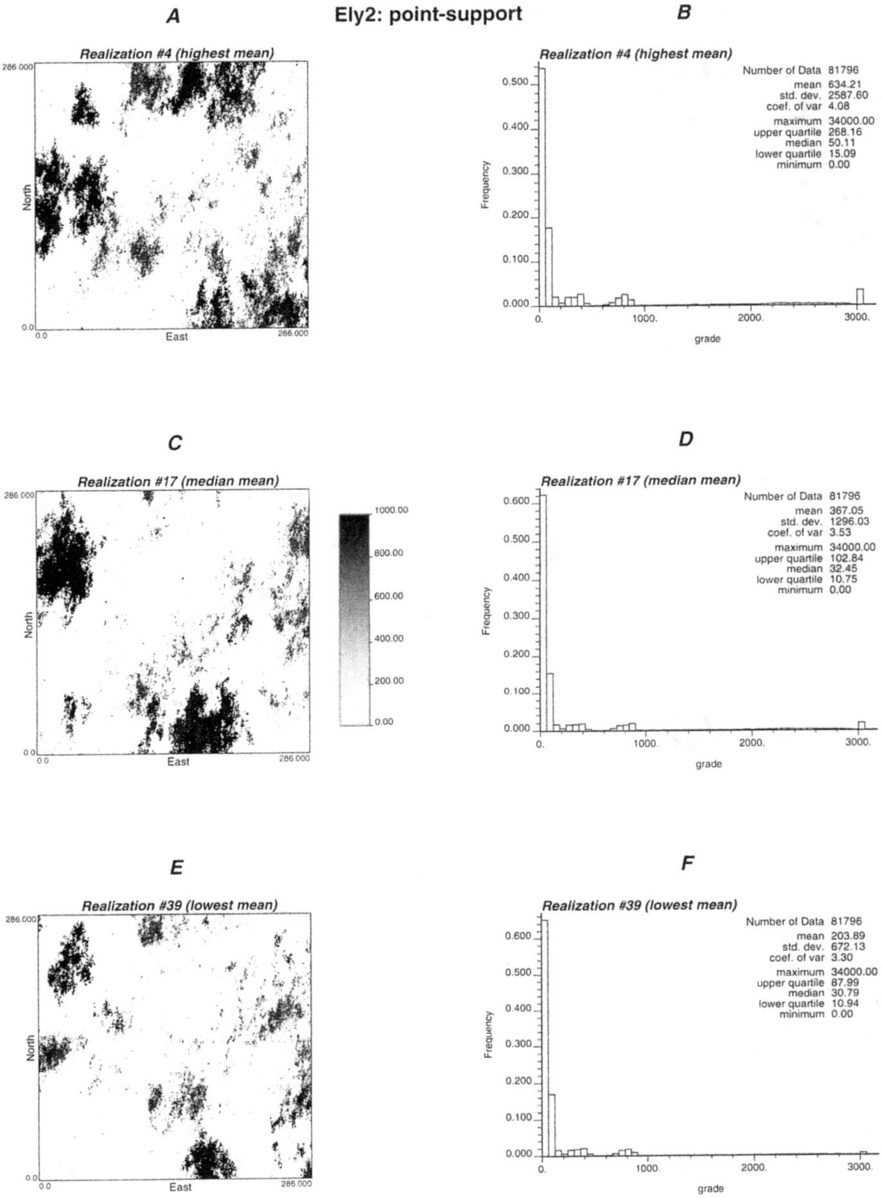

Figure V.29: Ely2: Three simulated point support grade realizations and their statistics.

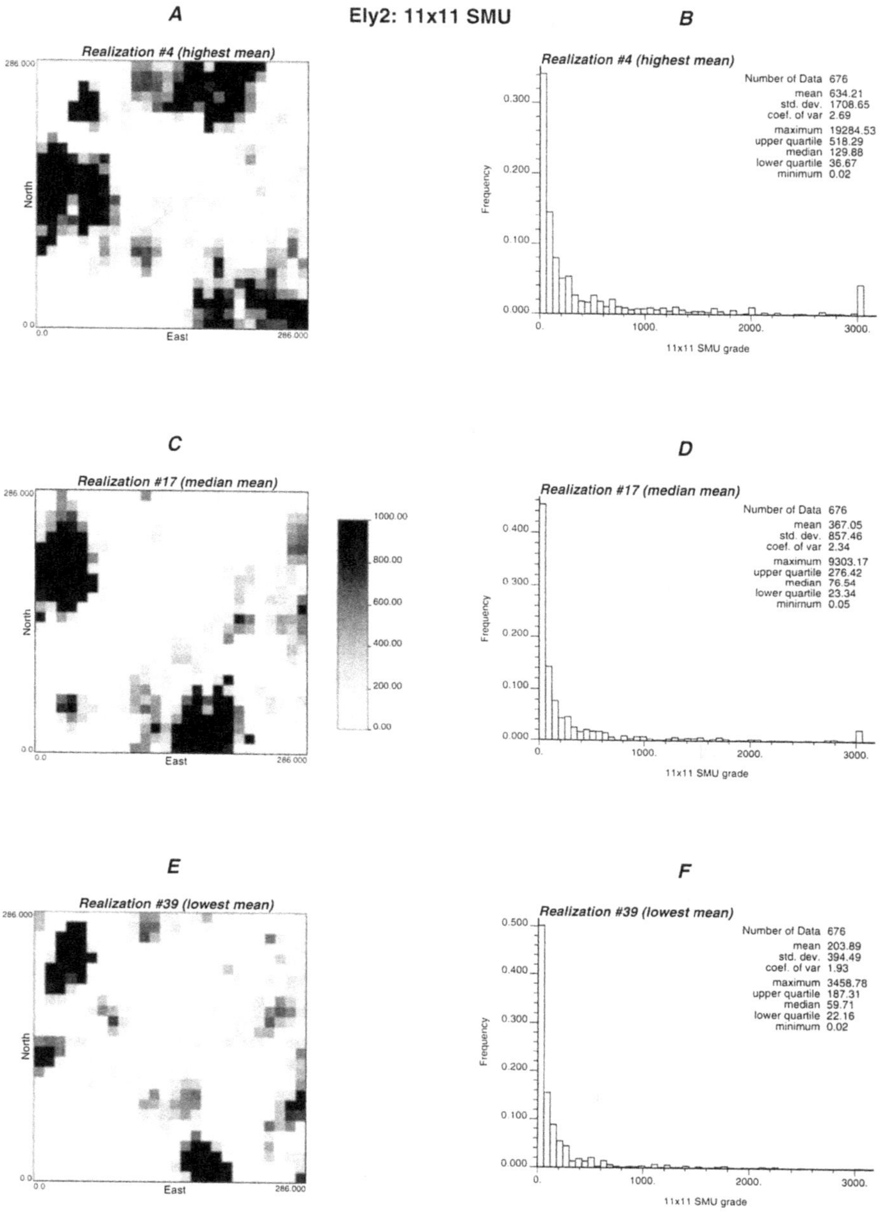

Figure V.30: Ely2: Three simulated 11×11 SMU grade realizations and their statistics.

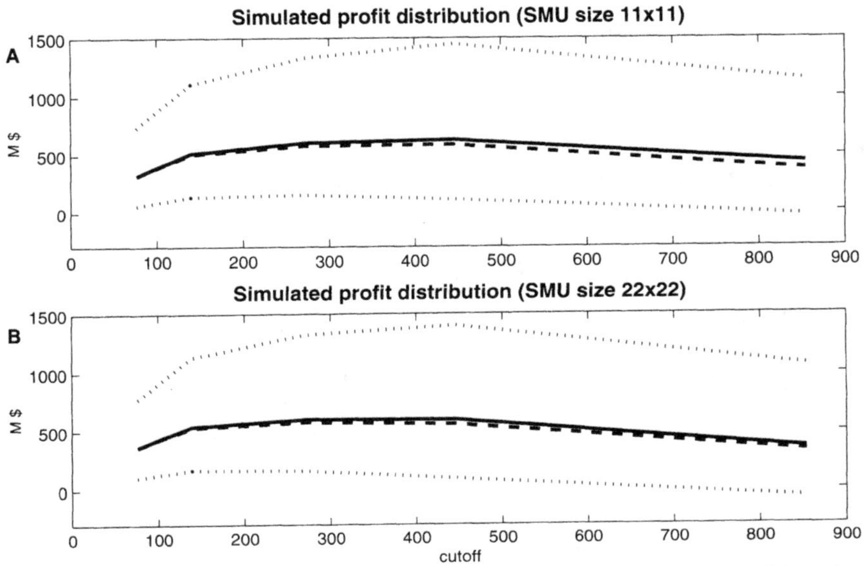

Figure V.31: Ely2 simulated profit under perfect selection for SMU size 11×11 (**A**) and 22×22 (**B**). **Dashed line:** mean simulated profit, **dotted lines:** 95% probability interval from distribution of simulated profit, **solid line:** "true" profit corresponding to perfect selection.

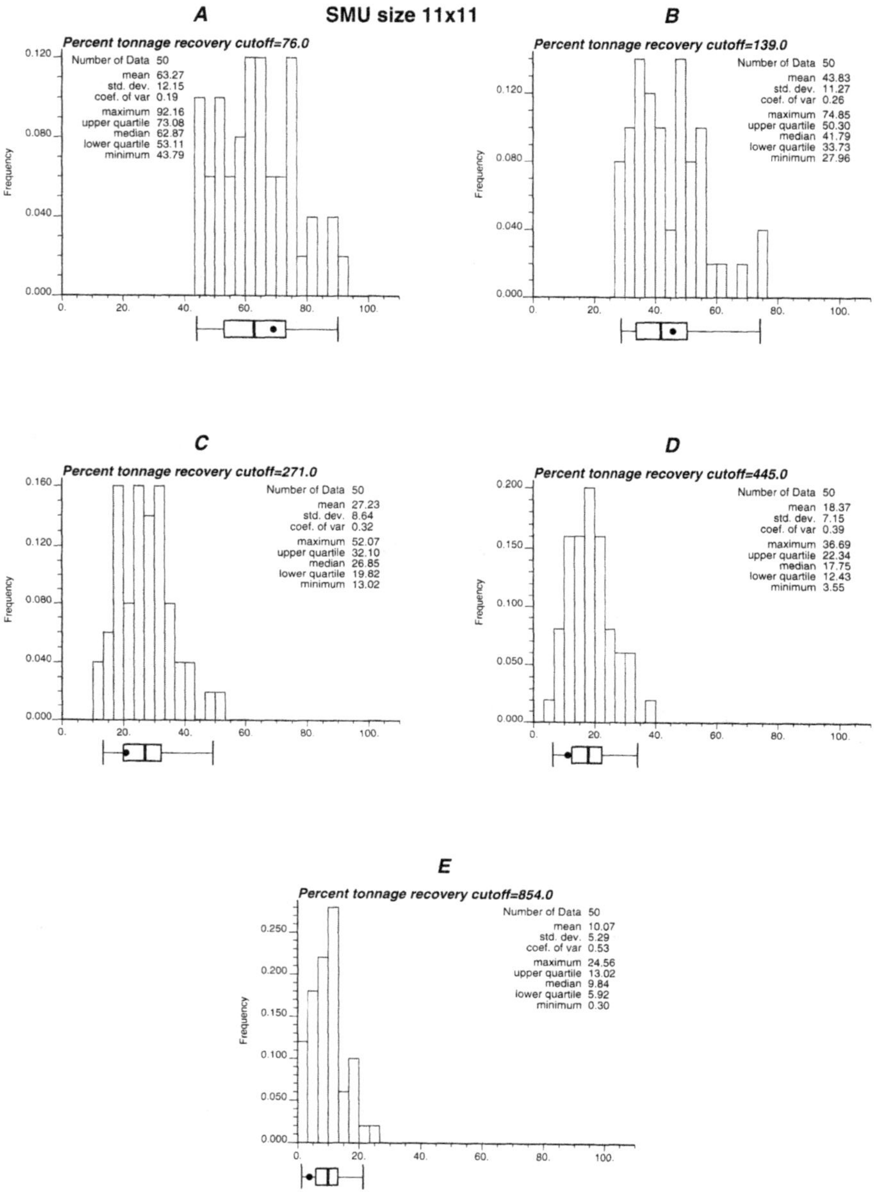

Figure V.32: Ely2: Simulated percent tonnage recovery under actual selection (SMU size: 11×11).

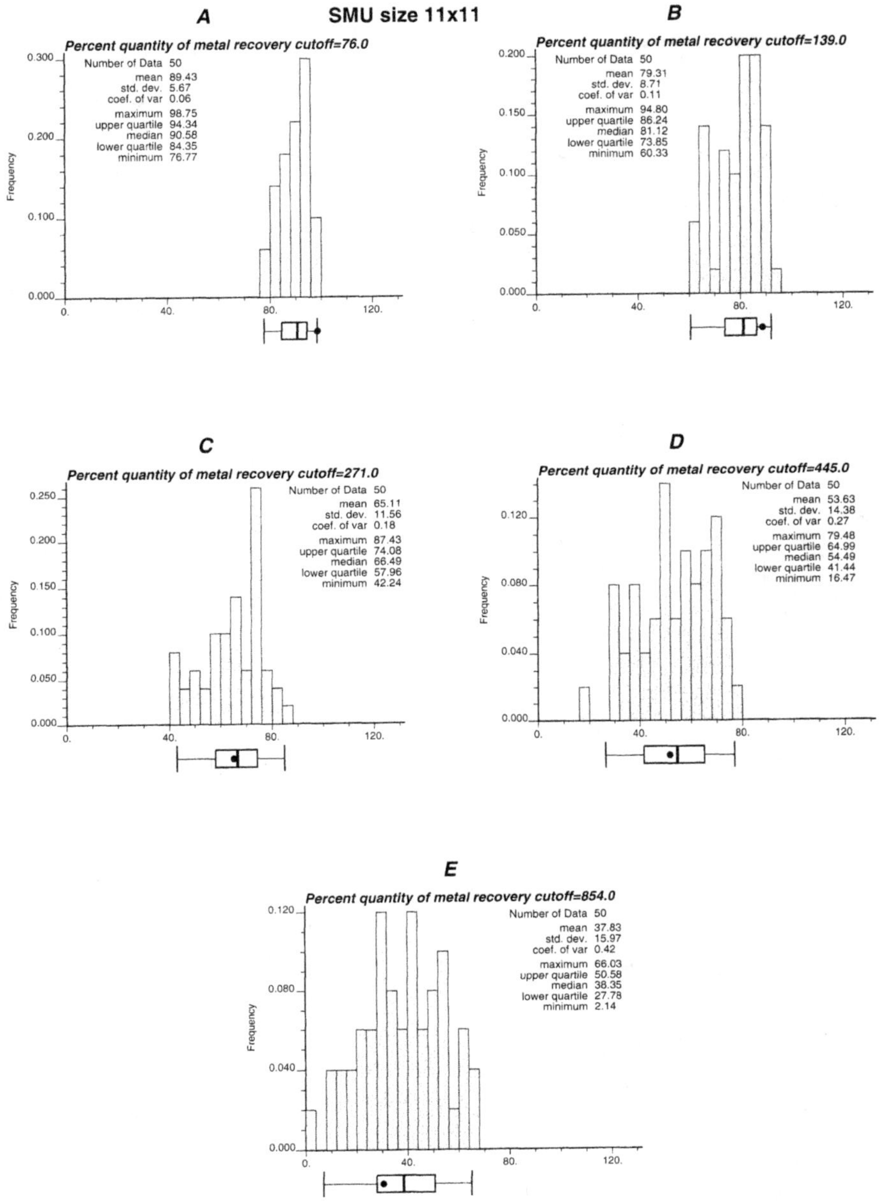

Figure V.33: Ely2: Simulated percent quantity of metal recovery under actual selection (SMU size: 11×11).

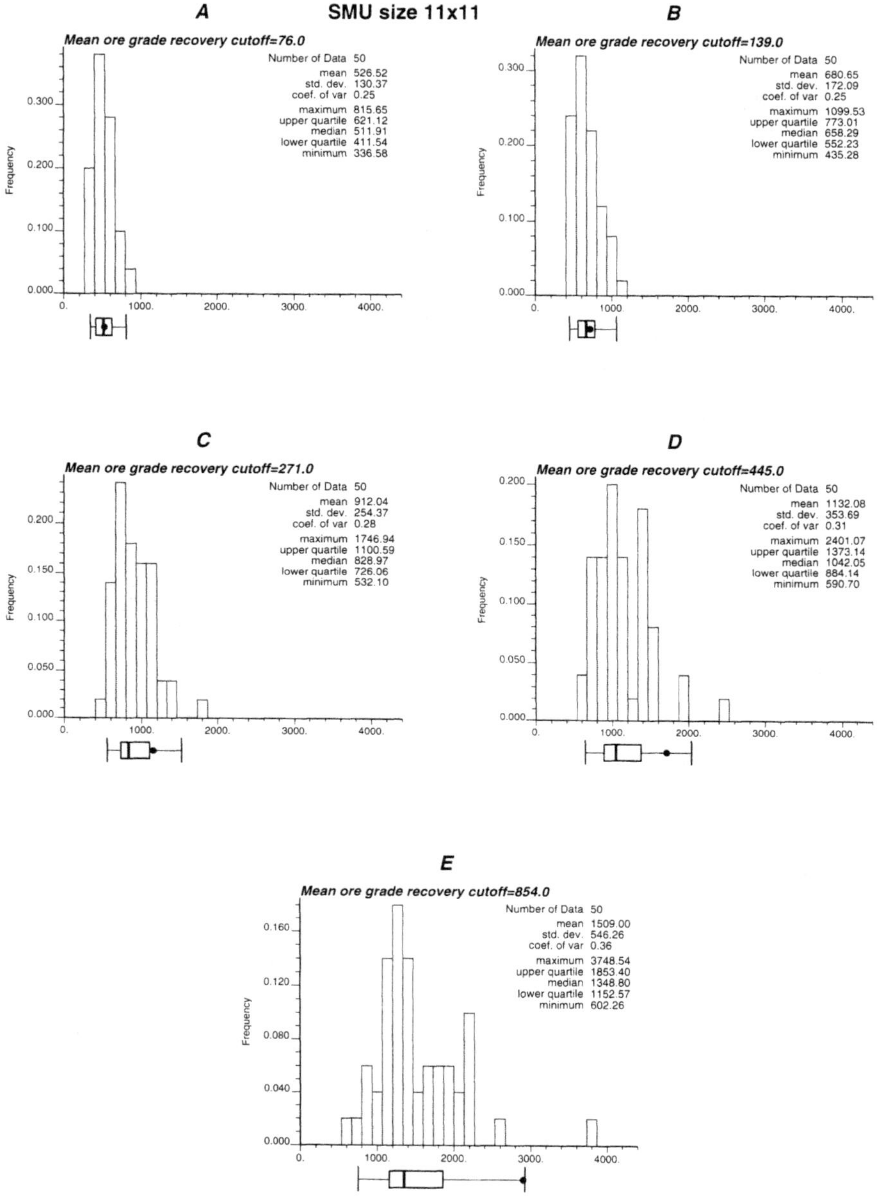

Figure V.34: Ely2: Simulated mean ore grade recovered under actual selection (SMU size: 11×11).

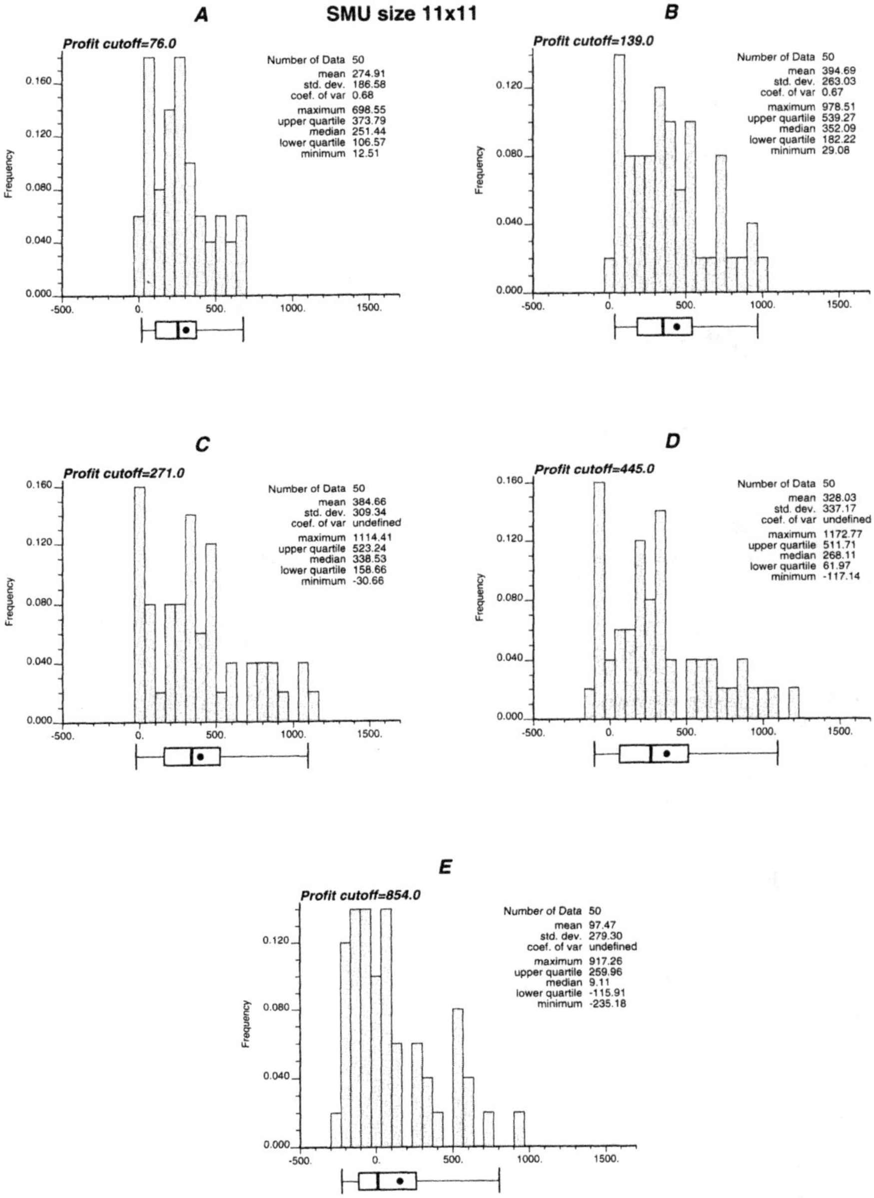

Figure V.35: Ely2: Simulated profit under actual selection (SMU size: 11 × 11).

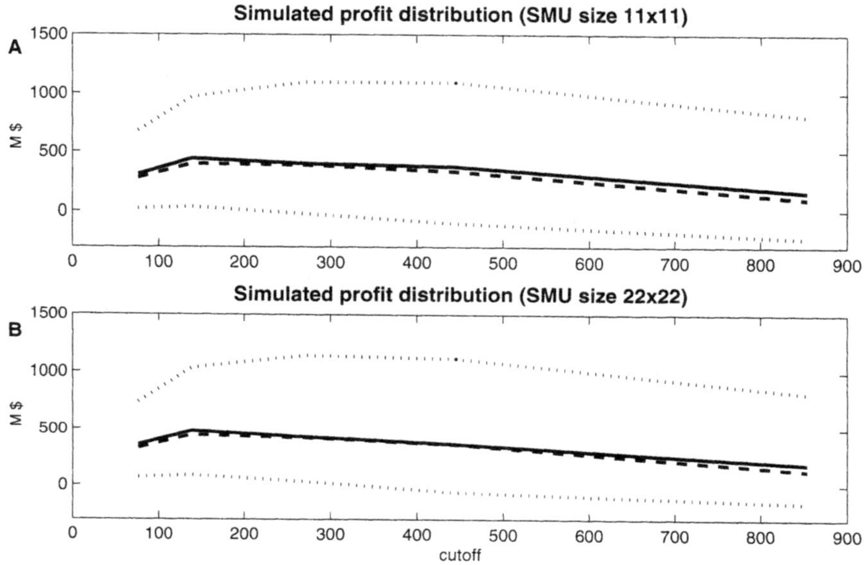

Figure V.36: Ely2 simulated profit under actual selection for SMU size 11×11 (**A**) and 22×22 (**B**). **Dashed line:** mean simulated profit, **dotted lines:** 95% probability interval from distribution of simulated profit, **solid line:** "true" profit corresponding to actual selection.

V.5 Summary

A simulation approach to the assessment of recoverable reserves should include several (L) realizations of the following:

- The spatial distribution of quasi-point support grade values $z^{(l)}(\mathbf{u})$ of the same support as, and conditioned to, the presently available grade data $z(\mathbf{u}_\alpha)$. These point support simulated values are then averaged into simulated SMU grade values $z_v^{(l)}(\mathbf{u})$.

- The spatial distribution of future selection data denoted $b^{(l)}(\mathbf{u}')$. These future selection data are typically of support and quality different from the presently available data; for example, blasthole data versus drillhole data, thus $b^{(l)}(\mathbf{u}_\alpha) \neq z(\mathbf{u}_\alpha)$. These simulated future data are then combined into simulated future SMU grade estimates $z_v^{**(l)}(\mathbf{u})$.

Simulation of perfect recovery (ignoring misclassification) consists of applying the cutoff grade z_c to the set of simulated true SMU grades $z_v^{(l)}(\mathbf{u})$ and retrieving the corresponding recovery functions.

Simulation of actual recovery consists of applying the cutoff grade z_c to the set of simulated future estimated values $z_v^{**(l)}(\mathbf{u})$. Once a SMU is simulated as ore, $z_v^{**(l)}(\mathbf{u}) > z_c$, it is the simulated true grade $z_v^{(l)}(\mathbf{u})$ that is sent to the mill; that is, used for simulation of the actual recovered quantity of metal and profit.

Although perfect recovery is never achieved in practice, its simulation is useful for evaluating the potential loss of profit resulting from misclassification at the time of mining. By simulating different qualities and densities of future selection data, the $b^{(l)}(\mathbf{u}')$ values, leading to different actual recoveries, one can simulate the impact of the information effect on profit. This is tantamount to a rigorous determination of the mining dilution factors widely used in practice to decrease estimates of ore grades and profits. Recall that this book does not address unpredictable, yet possibly major, dilutions resulting, for example, from heave of ore during blasting or difficulty to respect diglines at the time of mining.

Application of the proposed simulation approach to the first Ely reference data set resulted in reasonably accurate prediction of profit under perfect selection (recall Fig. V.10). The 95% probability interval, however, is very large, indicating that there is a substantial probability for "bad luck" that could result in either severe underestimation or overestimation of the reference true profit (recall also the final discussion of Section V.1).

Although any single simulated point grade realization does look significantly different from the reference point support map, once averaged over a SMU support, that realization becomes much more similar to the reference SMU support map. This is the saving grace of mining evaluation: Estimation errors or simulation deviations from the true reference tend to average out over SMU supports.

Prediction of recovery under actual selection over Ely was equally good, again with a large 95% probability interval (recall Fig. V.22). The score becomes slightly poorer when a poor variogram model, not reflecting the true grade anisotropy, was used (recall Fig. V.25).

Application of the proposed simulation approach to the second and more difficult Ely2 data set resulted in similarly good prediction, particularly for recovery under actual selection (recall Figs. V.31 and V.36). In this case, the reference actual profit was slightly underestimated by the mean simulated profit but remained well bracketed by the simulated 95% probability interval.

The Ely2 prediction results serve as a reminder that the flexibility of the simulation approach is no replacement for enough data of good quality at the time of prediction. Adequate-quality data at the present time of prediction allow for a reliable representation/simulation of those grade spatial structures that affect future ore recovery. Of equal importance is an approximate knowledge of the error characteristics of the future blasthole data, which dictate the future selection process. Any realistic attempt to successfully predict future recoverable reserves depends critically on an accurate prediction of the error expected in the future blasthole data and on an understanding of its complex relationship with the true (unknown) grades.

Chapter VI

Estimation of Recovery via Volume-Variance Correction

To evaluate the results obtained from the proposed simulation approach, we now proceed to estimation of the recovery functions using the traditional volume-variance correction (VVC) described in Chapter III. Remember, though, that the VVC approach does not account for the information effect; that is, the fact that SMUs are selected on the basis of future estimated values not on their true average grades. The VVC approach thus provides only estimates of recovery under perfect selection, not of actual recovery under imperfect selection.

Exactly the same prior information ($n = 30$ sample data and a variogram model) used for the simulation approach was provided to the VVC approach. The VVC procedure was applied to the first reference data set (Ely) with detailed results given, and then it was repeated on the second data set (Ely2) to evaluate its robustness.

A brief recall of the VVC procedure in the context of the Ely data set is given in Section VI.1. The prediction results using VVC before compositing; that is, before averaging of ccdfs within the panel, are given in Section VI.2. The results using VVC after compositing are given in Section VI.3. The robustness of these results with respect to the reference data set used is addressed in Section VI.4. The last section compares the VVC prediction to the prediction obtained from simulation.

VI.1 Volume-Variance Correction

The Ely data set presented in Sections IV.1 and IV.2 is used again for further analysis. Retaining the $n = 30$ sample data shown in Figure V.1A, median indicator kriging was performed to build the local conditional cumulative distribution functions (ccdfs) of the point support grades modeled as random variables $Z(\mathbf{u})$. These ccdfs are denoted as:

$$F(\mathbf{u}; z|(n)) = \text{Prob}\{Z(\mathbf{u}) \leq z|(n)\}, \quad \forall \mathbf{u} \in V \tag{VI.1}$$

Indicator kriging

Median IK was performed at all 676 nodes of the grid of size 11×11 discretizing the panel V. The single indicator semivariogram used was standardized to a unit sill and identified to the reference semivariogram $\gamma_Z(\mathbf{h})$ given in expression (IV.2). Recall that the sill of the indicator semivariogram model plays no role in the indicator kriging results. Thus:

$$\begin{aligned} \gamma_I(\mathbf{h}; z) &= \gamma_Z(\mathbf{h}), \quad \forall \text{ cutoff } z \\ &= 0.65\text{Exp}\sqrt{\left(\frac{h_1}{70}\right)^2 + \left(\frac{h_2}{35}\right)^2} + 0.35\text{Exp}\sqrt{\left(\frac{h_1}{75}\right)^2 + \left(\frac{h_2}{40}\right)^2} \end{aligned}$$

We could have adopted the γ_Z-model (V.1) inferred from the sole $n = 30$ sample data. We decided to retain the better anisotropic model above, because in a real-world situation geological interpretation might have indicated presence of anisotropy. From only 30 sample data, there is no hope in differentiating indicator variograms at the five different cutoffs, hence our decision to retain median indicator kriging with the same indicator variogram model for all cutoffs (recall the discussion in Section III.1).

The GSLIB program `ik3d` was used with the median indicator kriging option; the associated parameter file is given in Figure VI.1. For the Ely data set, the resulting maps of IK-derived probability values $1 - F(\mathbf{u}_j; z|(n))$ on the 11×11 grid ($j = 1, \ldots, 169$) are given in Figure VI.2. The five cutoff grade values z considered were the same retained for the simulation study as given in the list (IV.3). Remarkably, the 30 sample data suffice to detect the main high-grade zones; compare the IK-derived probability map at the median cutoff $z = 76$ (Fig. VI.2A) to the reference indicator map (Fig. IV.1A). The only high-grade zone missed is at the NE corner where there are no sample data (recall the sample location map of Fig. V.1A).

Volume-variance correction

The IK-derived probabilities, as displayed in Figure VI.2, relate to point support grades $z(\mathbf{u})$: They are models of the probability for point grade values to be greater than the cutoff. Such probabilities must be corrected for the SMU support effect by VVC. As discussed previously in Section III.2, there are two options:

- Option (1), "before compositing": VVC is applied to each of the $N = 676$ point ccdfs $F(\mathbf{u}_j; z|(n))$ discretizing the panel V. Only then are the N corrected ccdfs $F_v(\mathbf{u}_j; z|(n))$ averaged into a composite SMU ccdf $F^{(1)}_{v/V}(z|(n))$ over the panel V:

$$F^{(1)}_{v/V}(z|(n)) = \frac{1}{N}\sum_{j=1}^{N} F_v(\mathbf{u}_j; z|(n)) \tag{VI.2}$$

where

$$\begin{aligned} F_v(\mathbf{u}_j; z|(n)) &= \text{Prob}\{Z_v(\mathbf{u}_j) \leq z|(n)\} \\ &= \phi_{VVC}\left(v, F(\mathbf{u}; z|(n))\right) \end{aligned}$$

with $\phi_{VVC}(\cdot)$ designating the VVC applied to the SMU support v.

- Option (2), "after compositing": the N point ccdfs $F(\mathbf{u}_j; z|(n))$ are first averaged into a composite point ccdf $F_{\cdot/V}(z|(n))$ over the panel V; only then VVC is applied to correct for the SMU support v:

$$F^{(2)}_{v/V}(z|(n)) = \phi_{VVC}\left(v, F_{\cdot/V}(z|(n))\right) \tag{VI.3}$$

where

$$F_{\cdot/V}(z|(n)) = \frac{1}{N}\sum_{j=1}^{N} F(\mathbf{u}_j; z|(n))$$

Under both options, the result is a composite SMU ccdf within panel V:

$$F^{(\cdot)}_{v/V}(z|(n)) = \text{Prob}^{(\cdot)}_{v/V}\{Z_v(\mathbf{u}) \leq z|(n))\}, \ \forall\, \mathbf{u} \in V$$

which is an estimate (actually a model) of the probability that a SMU $v(\mathbf{u})$ taken at random within panel V has grade no greater than the cutoff value z. This probability is conditioned to the n sample data used.

The program for compositing (averaging) N distributions is trivial. The VVC correction $\phi_{VVC}(v, \cdot)$ is performed using GSLIB program `postik` and the indirect lognormal correction option. The associated parameter file used for option (1) "before compositing" is given in Figure VI.3. Program `postik` also provides the required recovery values defined on the SMU support ccdf $F^{(\cdot)}_{v/V}(z|(n))$.

In practice, there are many variants of the two previous implementations of VVC. This is not the place to test variants of an approach that we believe is better replaced by the simulation approach.

Variance reduction parameter

As discussed in Section III.3, the VVC procedure requires knowledge of the variance correction ratio f; that is, the ratio of the SMU-support variance to the point support variance. For this application, the ratio of the two conditional variances is approximated by the ratio of the corresponding non-conditional or dispersion variances, which can be calculated directly from the standardized γ_Z-semivariogram model [recall expression (III.10)]:

$$f = \frac{Var\{Z_v(\mathbf{u})|(n)\}}{Var\{Z(\mathbf{u})|(n)\}} \simeq \frac{D^2(v/\infty)}{D^2(\cdot/\infty)} = 1 - \bar{\gamma}_Z(v,v) \qquad \text{(VI.4)}$$

For the Ely data set, the variance correction ratio is:

$f = 0.71$ for the 11×11 SMU size, and

$f = 0.52$ for the 22×22 SMU size.

The mean semivariogram value $\bar{\gamma}_Z(v,v)$ was calculated using a 10×10 discretization of the SMU support v.

These f-ratio values calculated from the variogram model value $\bar{\gamma}_Z(v,v)$ can be compared to the actual ratio of the corresponding true variances obtained from the reference data set (see Fig. IV.5). The ratios of the true variances are:

$f^0 = (419)^2/(483)^2 = 0.75$ for the 11×11 SMU size, and

$f^0 = (357)^2/(483)^2 = 0.55$ for the 22×22 SMU size.

These actual f^0-values are higher than the previous calculated f-values (0.71 and 0.48). The calculated f-values thus overstate slightly the actual variance reduction caused by grade averaging, with too high a correction for the dilution associated with the support effect.

Permanence of distribution

VVC provides the corrected variances for the various ccdfs considered. The symmetrization of the probability distributions around their means calls for a permanence of distribution model. As discussed in Section III.3, it would be a severe error to use the affine correction model, which does not provide any symmetrization. We have adopted here the indirect lognormal correction algorithm (III.13), also available in the GSLIB program `postik`. Recall that this option would ensure permanence of lognormality were the initial point support distribution lognormal.

VI.2 Volume-Variance Correction Before Compositing: Results

As already mentioned in Section III.2, the advantage of performing VVC before compositing is that the ccdf of each SMU $v(\mathbf{u})$ within the panel V is available; indeed from relation (VI.2):

$$F_v(\mathbf{u}; z|(n)) = \text{Prob}\{Z_v(\mathbf{u}) \leq z|(n)\} = \phi_{VVC}(v, F(\mathbf{u}; z|(n)))$$

This ccdf represents the probability for the specific SMU $v(\mathbf{u})$ centered at location $\mathbf{u}$ to be no greater than the cutoff grade z.

The complement probability $1 - F_v(\mathbf{u}; z|(n))$ represents the probability that the specific SMU $v(\mathbf{u})$ be selected at cutoff z; it is not an estimate of the future tonnage recovered, as that tonnage recovery will be all (1) or nothing (0) per the definition of support $v(\mathbf{u})$ being the minimum selection unit. To avoid any possible confusion we will not use the tonnage notation for that probability:

$$1 - F_v(\mathbf{u}; z|(n)) = \text{Prob}\{Z_v(\mathbf{u}) > z|(n)\}$$

There is one such probability value for each of the 676 SMUs of size 11×11 within panel V and for each cutoff grade z. These probability values can thus be mapped. Figure VI.4 gives five such probability maps for SMU size 11×11 and for the five cutoff grades z retained. Compare these probability maps with the reference indicator maps of Figure IV.7.

Consider now applying the probability threshold $p_0 = 0.537$ to the probability map of Figure VI.4A; that is, declare a SMU $v(\mathbf{u})$ as above the cutoff $z = 76$ if $1 - F_v(\mathbf{u}; z|(n))$ exceeds p_0. The corresponding indicator is defined as:

$$k_v(\mathbf{u}; 76) = \begin{cases} 1 & \text{if } 1 - F_v(\mathbf{u}; 76|(n)) > p_0 = 0.537 \\ 0 & \text{if not} \end{cases} \tag{VI.5}$$

Figure VI.5B gives the map of the 676 binary indicators $k_v(\mathbf{u}; 76)$ within panel V; Figure VI.5A recalls the reference indicator map for the same cutoff $z = 76$. The two maps are reasonably similar. The probability threshold $p_0 = 0.537$ was here selected such that the proportion of SMUs declared as profitable (black pixels) in the two maps is approximately the same ($\simeq 0.5$). In a real study situation, p_0 could have been determined from a test bench where both Figures VI.5A and VI.5B would be available or simulated (this is a challenge for new research in mining geostatistics). For more information on the concept of replacing cutoff grades (z) by related probability thresholds (p_0), the reader is referred to the pioneering works of Glacken (1997) and Kyriakidis (1997).

From the ccdf $F_v(\mathbf{u}; z|(n))$ specific to each SMU $v(\mathbf{u})$, one could retrieve the following truncated statistics above cutoff z:

$$\begin{aligned} qq_v(\mathbf{u}; z) &= \int_z^\infty y dF_v(\mathbf{u}; y|(n)) \\ mm_v(\mathbf{u}; z) &= E\{Z_v(\mathbf{u})|Z_v(\mathbf{u}) > z, (n)\} \\ &= qq_v(\mathbf{u}; z)/[1 - F_v(\mathbf{u}; z|(n))] \end{aligned} \tag{VI.6}$$

Beware that these truncated statistics are *not* direct estimates of the quantity of metal and mean ore grade of SMU $v(\mathbf{u})$, they are conditional to the SMU $v(\mathbf{u})$ passing the cutoff grade z; that is, conditional to $Z_v(\mathbf{u}) > z$. This is the reason for the different notations $qq_v(\cdot)$ and $mm_v(\cdot)$ instead of $q_v(\cdot)$ and $m_v(\cdot)$.

If we suggest plotting the probability maps $\{1 - F_v(\mathbf{u}; z|(n)), \mathbf{u} \in V\}$ (see Fig. VI.4), we do not recommend plotting similar maps for the truncated statistics (VI.6): These can be too easily mistaken for estimates of the in situ SMU grades.

Global recovery estimates
The recovery estimates within panel V can be calculated directly from the composite SMU ccdf $F^{(1)}_{v/V}(z|(n))$ as defined in expression (VI.2); that is, as obtained from VVC before compositing. Equivalently, they can be calculated by adding the expected recovery values obtained from each of the N ccdfs $F_v(\mathbf{u}_j; z|(n))$ specific to each of the N SMUs $v(\mathbf{u}_j)$ within panel V. Recalling expressions (III.15), the corresponding recovery functions are:

- Ore tonnage:

$$t^{*(1)}_{v/V}(z) = N\left[1 - F^{(1)}_{v/V}(z|(n))\right] = \sum_{j=1}^{N} [1 - F_v(\mathbf{u}_j; z|(n))] \qquad \text{(VI.7)}$$

- Quantity of metal:

$$\begin{aligned} q^{*(1)}_{v/V}(z) &= N\int_z^{\infty} y dF^{(1)}_{v/V}(y|(n)) \\ &= \sum_{j=1}^{N}\int_z^{\infty} y dF_v(\mathbf{u}_j; y|(n)) = \sum_{j=1}^{N} qq_v(\mathbf{u}_j; z) \end{aligned}$$

- Mean ore grade:

$$m^{*(1)}_{v/V}(z) = q^{*(1)}_{v/V}(z)/t^{*(1)}_{v/V}(z) \neq \frac{1}{N}\sum_{j=1}^{N} mm_v(\mathbf{u}_j; z)$$

The estimated profit is a known function of the estimated ore tonnage $t^{*(1)}_{v/V}(z)$ and quantity of metal $q^{*(1)}_{v/V}(z)$ recovered. Here we used the profit function (IV.4).

Table VI.1 gives the recovery values (VI.7) within panel V for SMU sizes 11×11 and 22×22. Figure VI.6 gives the corresponding recovery curves as function of the cutoff grade z. Tonnage and quantity of metal are given in percent recovery of the total estimated at zero cutoff:

$$\begin{aligned} \%\ \text{tonnage} &= t^{*(1)}_{v/V}(z)/N \qquad \text{(VI.8)} \\ \%\ \text{quantity of metal} &= q^{*(1)}_{v/V}(z)/q^{*}_{v/V}(0) \end{aligned}$$

The mean ore grade is given in grade units, and the profit is expressed in $M\$$.

The total quantity of metal (at cutoff $z = 0$) is:

$$q^*_{v/V}(0) = 81796 \times m_{IK}(0)$$

where $m_{IK}(0) = 267$ denotes the median IK-derived overall mean grade, here obtained as an output of GSLIB programs **ik3d** and **postik**; the parameter files of these programs are given in Figures VI.1 and VI.3 (see also Deutsch and Journel, 1998, p. 103–106 and 236–238).

From Table VI.1, the VVC approach option (1) "before compositing" gives recovery functions reasonably close to the reference values achieved under perfect selection (recall the corresponding results given in Table IV.1 of Section IV.2). Recall, however, that these results do not relate to what a mining operation would achieve in practice; that is, recovery under imperfect selection.

11 × 11 SMU				
Cutoff (z)	% Tonnage	% Q. of Metal	Mean Grade	Profit
76	54.4	93.2	457	139
139	39.7	87.8	591	240
271	34.2	83.9	656	230
445	18.4	62.6	909	149
854	7.1	35.7	1350	-43
22 × 22 SMU				
Cutoff (z)	% Tonnage	% Q. of Metal	Mean Grade	Profit
76	60.9	94.6	415	188
139	41.6	87.5	562	281
271	34.5	82.6	638	265
445	18.7	60.9	871	175
854	6.4	32.4	1345	-8

Table VI.1: Recovery functions (Ely data set) for SMU sizes 11 × 11 and 22 × 22, estimated via indicator kriging and VVC before compositing. % tonnage refers to total tons = 286 × 286 = 81796 unit tons (at zero cutoff), % quantity of metal refers to total $q^*_{v/V}(0) = 81796 \times m_{IK}(0)$ unit tons × grade unit, where $m_{IK}(0)$ denotes the IK-derived overall mean grade ($m_{IK}(0) = 267$ for both 11 × 11 and 22 × 22 SMU size); profit is given in $M\$$.

```
                          Parameters for IK3D
                          *******************

START OF PARAMETERS:
1                                  -1=continuous(cdf), 0=categorical(pdf)
0                                  -option: 0=grid, 1=cross, 2=jackknife
jack.dat                           -file with jackknife data
1   2   0    3                     -   columns for X,Y,Z,vr
5                                  -number thresholds/categories
 76  139  271  445  854            -   thresholds / categories
0.5  0.6  0.7  0.8  0.9            -   global cdf / pdf
sample.dat                         -file with data
1   2   0    3                     -   columns for X,Y,Z and variable
nodata.dat                         -file with soft indicator input
1   2   0    3  4   5   6          -   columns for X,Y,Z and indicators
-1.0e21    1.0e21                  -trimming limits
0                                  -debugging level: 0,1,2,3
ik3d.dbg                           -file for debugging output
oik11.out                          -file for kriging output
26    5.5     11.0                 -nx,xmn,xsiz
26    5.5     11.0                 -ny,ymn,ysiz
1     0.0     5.0                  -nz,zmn,zsiz
4    16                            -min, max data for kriging
300.0 300.0 0.0                    -maximum search radii
  0.0  0.0  0.0                    -angles for search ellipsoid
0                                  -max per octant (0-> not used)
1    271                           -0=full IK, 1=Median IK(threshold num)
1                                  -0=SK, 1=OK
2      0.00                        -One   nst, nugget effect
2      0.65 10.0    0.0    0.0     -it,cc,ang1,ang2,ang3
            70.    35.0   10.0     -a_hmax, a_hmin, a_vert
2      0.35 10.0    0.0    0.0     -it,cc,ang1,ang2,ang3
            75.0   40.0   10.0      -a_hmax, a_hmin, a_vert
```

Figure VI.1: **ik3d** parameter file for indicator kriging on a 11×11 grid.

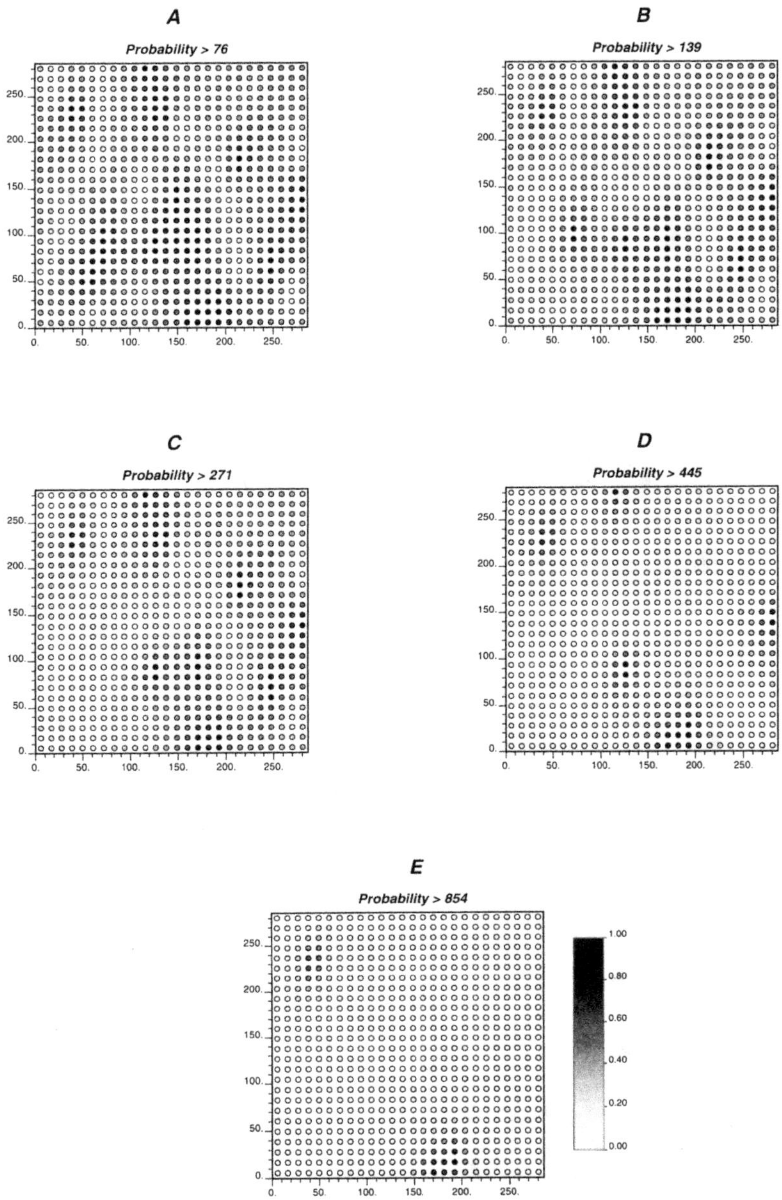

Figure VI.2: Grayscale maps of point support probability values $1 - F(\mathbf{u}; z|(n))$ defined on a 11×11 grid, estimated via indicator kriging using the $n = 30$ sample data (Ely data set).

```
                  Parameters for POSTIK
                  *********************

START OF PARAMETERS:
oik11.out                             -file with IK3D output (continuous)
postoik11cutoff1.out                  -file for output
2  75.84                              -output option, output parameter
5                                     -number of thresholds
75.84 139.31 271.02 444.59 854.16 -the thresholds
1   2      0.71                       -volume support?, type, varred
sample.dat                            -file with global distribution
3   0    -1.0   1.0e21                -   ivr,  iwt,  tmin,  tmax
0.0     7000.0                        -minimum and maximum Z value
1   1.5                               -lower tail: option, parameter
1   1.0                               -middle     : option, parameter
4   2.5                               -upper tail: option, parameter
100                                   -maximum discretization
```

Figure VI.3: **postik** parameter file for performing the volume-variance correction.

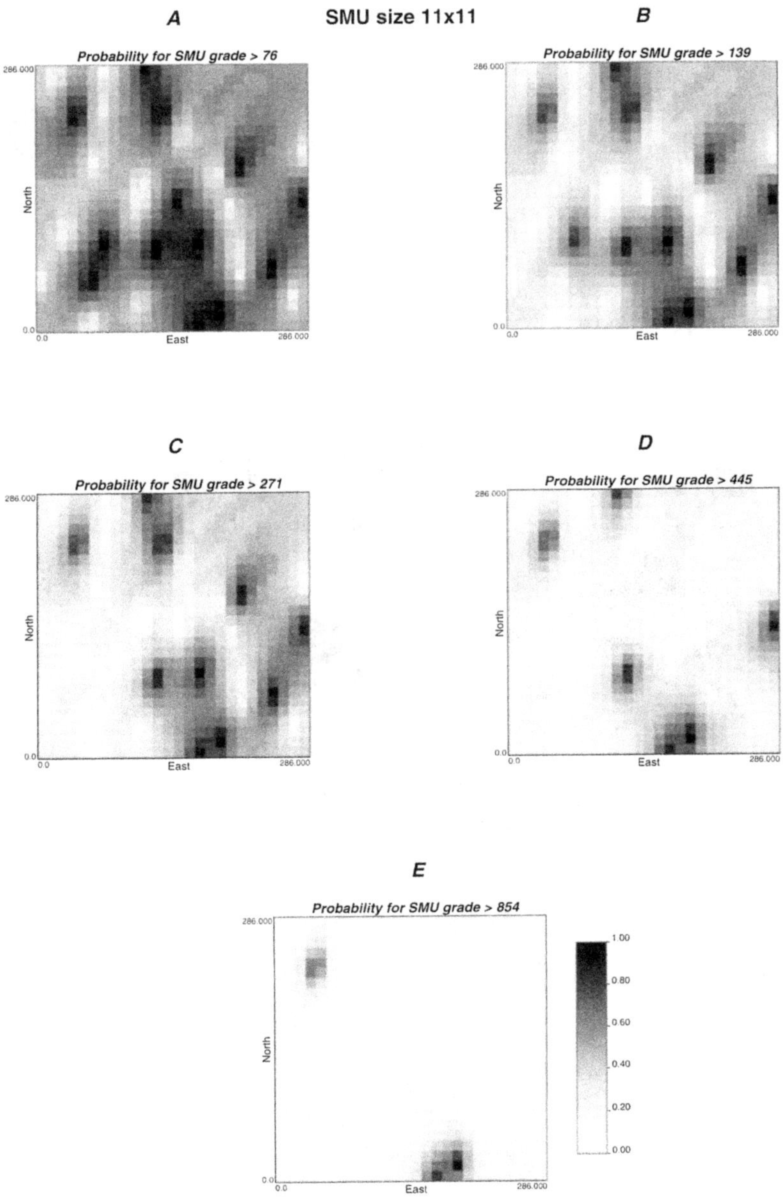

Figure VI.4: Grayscale maps of SMU probability values $1 - F_v(\mathbf{u}; z|(n))$, derived via VVC from the point support probabilities (Ely data set).

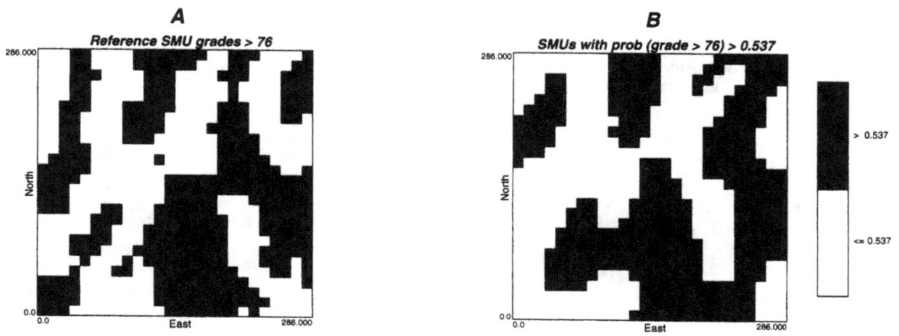

Figure VI.5: Indicators of 11 × 11 SMU grades above cutoff $z = 76$. **A:** reference indicators under perfect recovery, **B:** indicators obtained by applying the threshold $p_0 = 0.537$ to the local ccdf values $\text{Prob}\{Z_v(\mathbf{u}) > 76|(n)\}$, as determined by VVC.

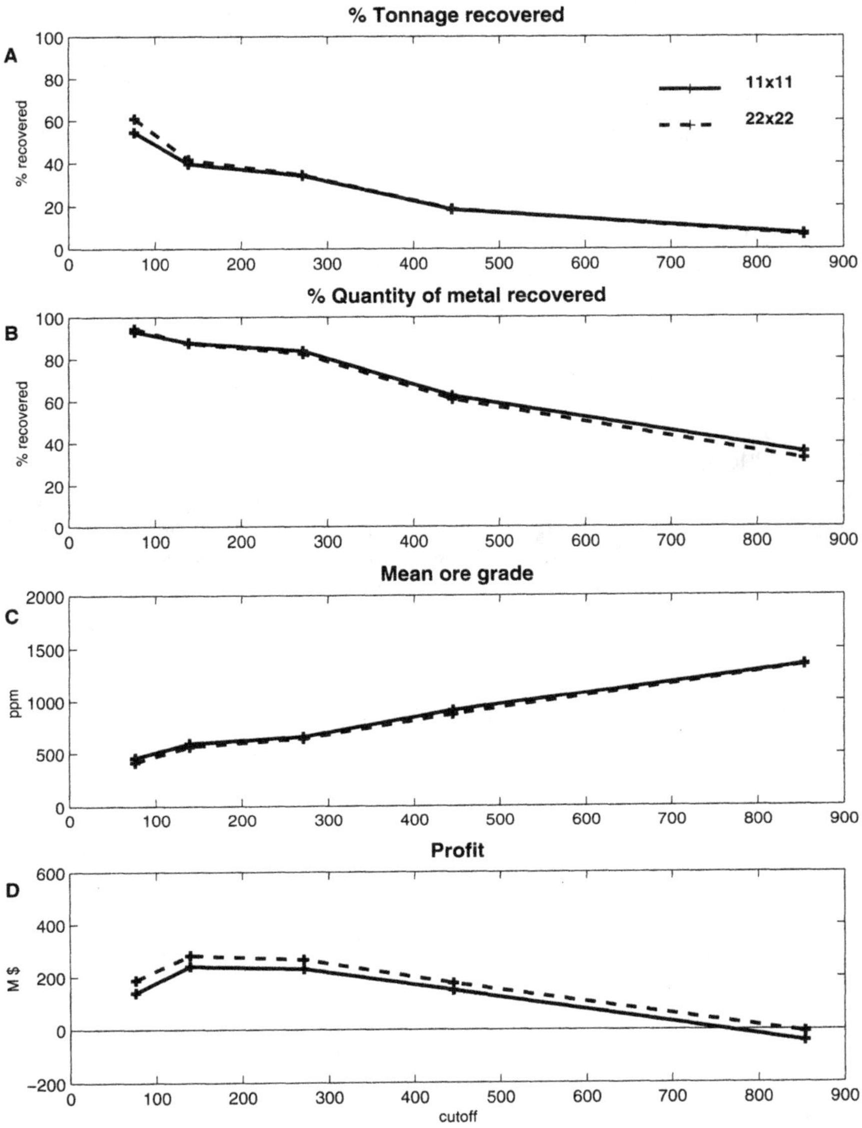

Figure VI.6: Recovery functions (Ely data set) for SMU sizes 11×11 and 22×22, estimated via indicator kriging and VVC before compositing. **A:** percent tonnage recovered (total tonnage is for zero cutoff), **B:** percent quantity of metal recovered, **C:** mean ore grade, **D:** profit.

VI.3 Volume-Variance Correction After Compositing: Results

When the VVC is performed after compositing, the only ccdf defined on SMU support is the composite SMU ccdf $F^{(2)}_{v/V}(z|(n))$ as defined in expression (VI.3). There is only one such ccdf per panel V, and hence no resolution within V is provided.

Global recovery estimates

The recovery estimates within panel V are obtained directly from the composite SMU ccdf $F^{(2)}_{v/V}(z|(n))$. The expressions are similar to (VI.7); it suffices to replace the superscript (1) by the superscript (2) corresponding to VVC after compositing. Table VI.2 gives the recovery values within panel V for SMU sizes 11×11 and 22×22, calculated via VVC after compositing. Figure VI.7 gives the corresponding recovery curves as function of the cutoff grade z. Recall, from (VI.8), that tonnage and quantity of metal are given in percent of the total estimated at zero cutoff. From Table VI.2, it appears that the VVC approach option (2) "after compositing" also gives recovery functions reasonably close to the reference values achieved under perfect recovery (recall the corresponding results given in Table IV.1). In general the two VVC options (1) and (2) give roughly similar results (compare Table VI.1 and Table VI.2).

11 × 11 SMU				
Cutoff (z)	% Tonnage	% Q. of Metal	Mean Grade	Profit
76	55.1	92.8	450	187
139	40.0	87.3	583	285
271	34.3	83.8	651	277
445	18.6	61.1	879	180
854	6.8	33.1	1309	-12

22 × 22 SMU				
Cutoff (z)	% Tonnage	% Q. of Metal	Mean Grade	Profit
76	62.0	94.1	406	184
139	42.3	87.4	552	278
271	34.9	81.7	625	254
445	18.9	59.8	842	158
854	5.9	29.0	1310	-29

Table VI.2: Recovery functions (Ely data set) for SMU sizes 11×11 and 22×22 estimated via indicator kriging and VVC after compositing. % tonnage refers to total tons $= 286 \times 286 = 81796$ unit tons (at zero cutoff), % quantity of metal refers to total $q^*_{v/V}(0) = 81796 \times m_{IK}(0)$ unit tons × grade unit, where $m_{IK}(0)$ denotes the IK-derived overall mean grade ($m_{IK}(0) = 267$ for both 11×11 and 22×22 SMU size); profit is given in $M\$$.

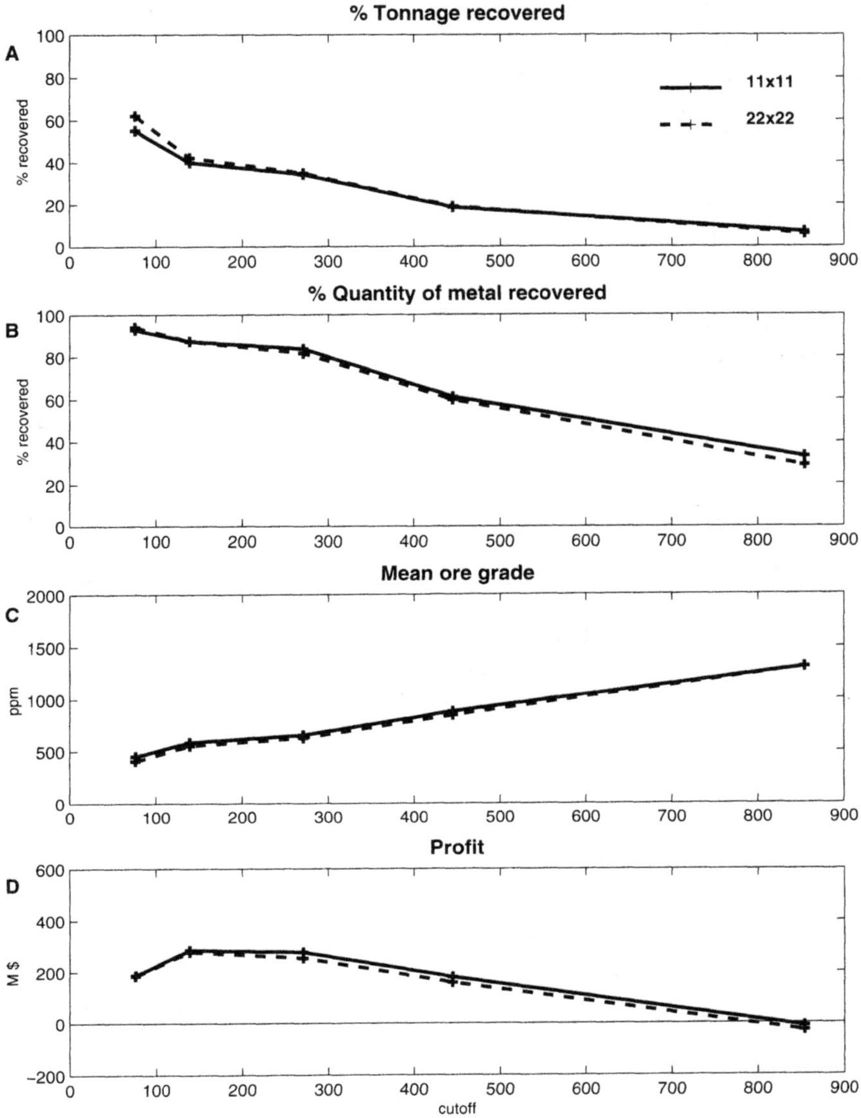

Figure VI.7: Recovery functions (Ely data set) for SMU sizes 11×11 and 22×22, estimated via indicator kriging and VVC after compositing. **A:** percent tonnage recovered (total tonnage is for zero cutoff), **B:** percent quantity of metal recovered, **C:** mean ore grade, **D:** profit.

VI.4 Comparison with Simulation Results

Recall that the VVC approach can provide only estimates of recovery under perfect selection, *not* estimates of actual recovery under imperfect selection. The simulation approach can provide estimates of both types of recovery with, in addition, an assessment of the related uncertainty.

Figure VI.8 overlays the simulated predictions of profit under perfect selection (as shown in Fig. V.10) with the profit curves estimated by VVC before compositing (taken from Fig. VI.6D). For the 22 × 22 SMU size, the predictions given by the VVC and the mean simulated value are equally good. For the 11 × 11 SMU size, the simulation appears slightly more accurate.

The critical recovery functions to be predicted, however, are those corresponding to imperfect selection: At the time of mining and actual selection there will be necessarily some misclassification (recall Fig. I.3). To make this point clear, we plotted in Figure VI.9 the reference profit derived from actual selection based on perfect-quality future blasthole data (taken from Table V.1) versus the corresponding prediction provided by simulation (taken from Fig. V.22) and versus the VVC-based prediction (under perfect selection, taken from Fig. VI.6D). Were the information effect (misclassification) ignored, as is too often done in the practice of recoverable reserves estimation, the VVC-based predictions may have been retained, resulting in a slight overestimation of the reference actual profit. For the 11 × 11 SMU size, it appears that the VVC underestimation of profit under perfect selection (see Fig. VI.8A) matches accidentally the profit loss obtained from actual selection; this leads to the (accidental) good VVC score shown in Figure VI.9A.

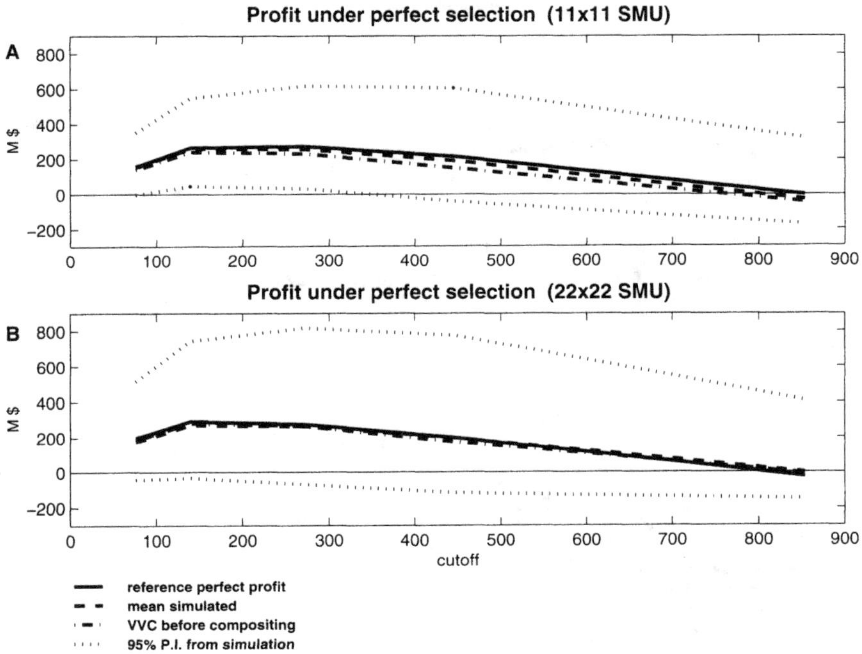

Figure VI.8: Profit under perfect selection (Ely data set) for SMU size 11 × 11 (**A**) and 22 × 22 (**B**). **Solid line:** true perfect profit, **dashed line:** mean simulated profit, **dash-dotted line:** profit predicted using VVC before compositing, **dotted lines:** 95% probability intervals from the distribution of simulated profits.

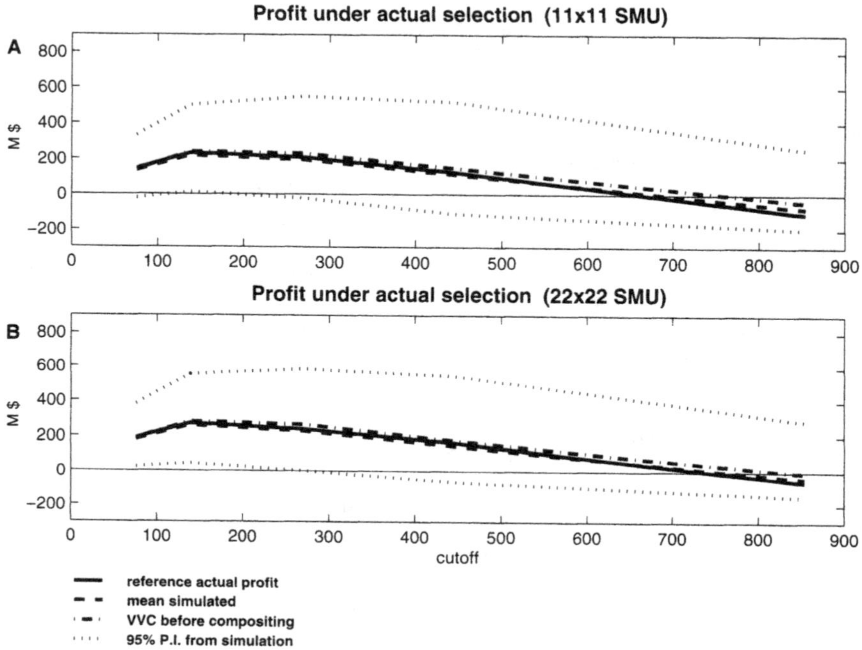

Figure VI.9: Profit under actual selection based on perfect-quality future blastholes (Ely data set) for SMU size 11 × 11 (**A**) and 22 × 22 (**B**). **Solid line:** true actual profit, **dashed line:** mean simulated profit, **dash-dotted line:** profit predicted using VVC before compositing (assuming perfect selection), **dotted lines:** 95% probability intervals from the distribution of simulated profits.

VI.5 Volume-Variance Correction: Results for Ely2

Under exactly the same conditions as for Ely, starting from $n = 30$ sample data, the VVC procedure was repeated for the Ely2 data set. Full (not median) indicator kriging was performed using the variogram model (V.7) for the first four cutoffs, and the variogram model (V.6) for the last cutoff $z = 854$. Because the VVC approach is not the subject of this book, only the final results are given in Table VI.3 and Figures VI.10 and VI.11. The estimates of recovery under perfect selection were found to be similar, whether VVC is performed before or after compositing.

Figure VI.12 overlays the simulated predictions of profit under perfect selection (as shown in Fig. V.31) with the profit curves estimated by VVC before compositing. It appears that the VVC-derived profit systematically underestimates the reference profit curve, especially for the 11×11 SMU size. The mean simulated profit approximates more closely the reference profit, while also providing uncertainty intervals.

11 × 11 SMU				
Cutoff (z)	% Tonnage	% Q. of Metal	Mean Grade	Profit
76	52.1	94.5	615	259
139	39.5	90.9	780	421
271	34.8	88.3	861	471
445	18.6	71.1	1294	447
854	7.6	50.6	2249	278
22 × 22 SMU				
Cutoff (z)	% Tonnage	% Q. of Metal	Mean Grade	Profit
76	55.7	95.2	579	309
139	40.3	90.8	764	468
271	34.9	87.9	853	515
445	18.9	70.7	1269	487
854	7.5	49.8	2246	321

Table VI.3: Recovery functions (Ely2 data set) for SMU sizes 11×11 and 22×22, estimated via indicator kriging and VVC before compositing. % tonnage refers to total tons $= 286 \times 286 = 81796$ unit tons (at zero cutoff), % quantity of metal refers to total $q^*_{v/V}(0) = 81796 \times m_{IK}(0)$ unit tons $\times$ grade unit, where $m_{IK}(0)$ denotes the IK-derived overall mean grade ($m_{IK}(0) = 339$ for both 11×11 and 22×22 SMU size); profit is given in $M\$$.

Figure VI.13 overlays the simulated predictions of profit under actual selection (as shown in Fig. V.36) with the profit curves estimated by VVC before compositing. For this more difficult second Ely2 data set, the simulation approach appears clearly superior to the VVC approach: the VVC-derived profit curve systematically overestimates the reference profit curve, whereas the mean simulated profit closely approximates that reference curve. Unlike the VVC approach, which cannot assess uncertainty in predicted profit, the simulation approach provides a richer input for risk-conscious mine planning via the realistic (albeit alarmingly large) probability intervals of Figure VI.13.

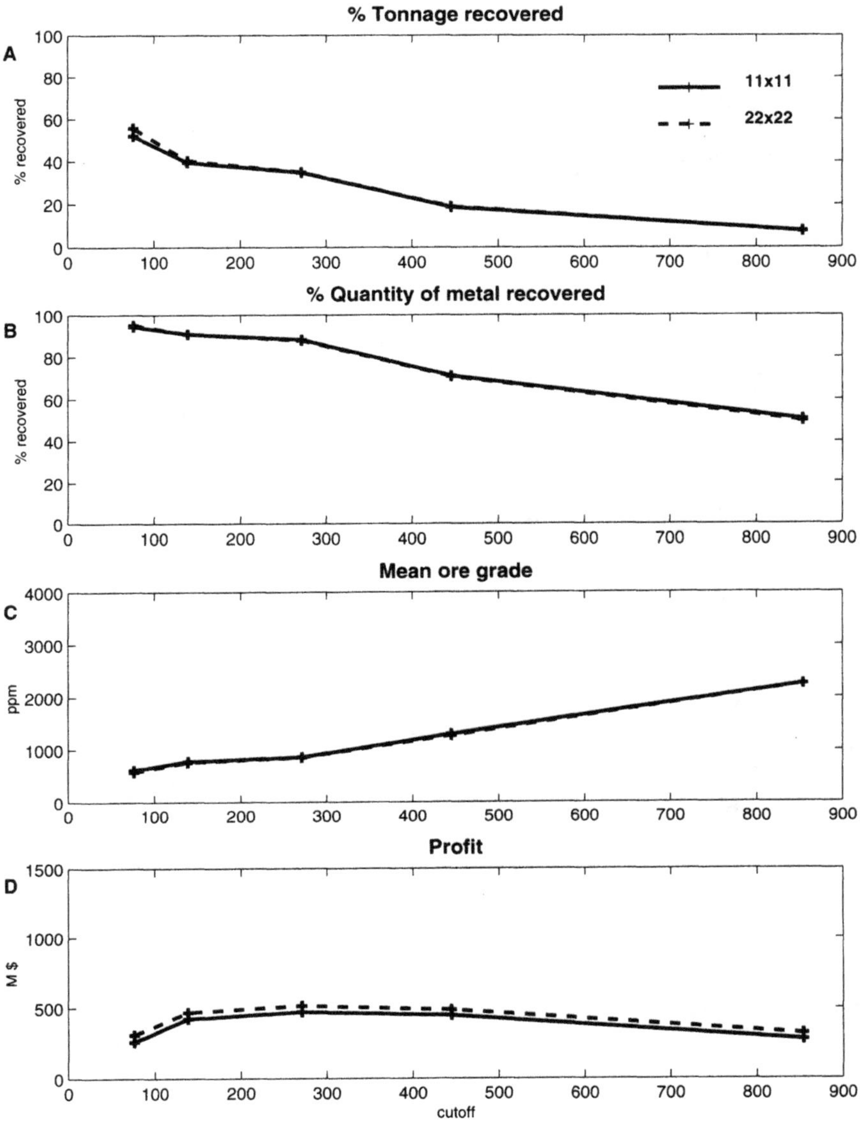

Figure VI.10: Recovery functions (Ely2 data set) for SMU sizes 11×11 and 22×22, estimated via indicator kriging and VVC before compositing. **A:** percent tonnage recovered (total tonnage is for zero cutoff), **B:** percent quantity of metal recovered, **C:** mean ore grade, **D:** profit.

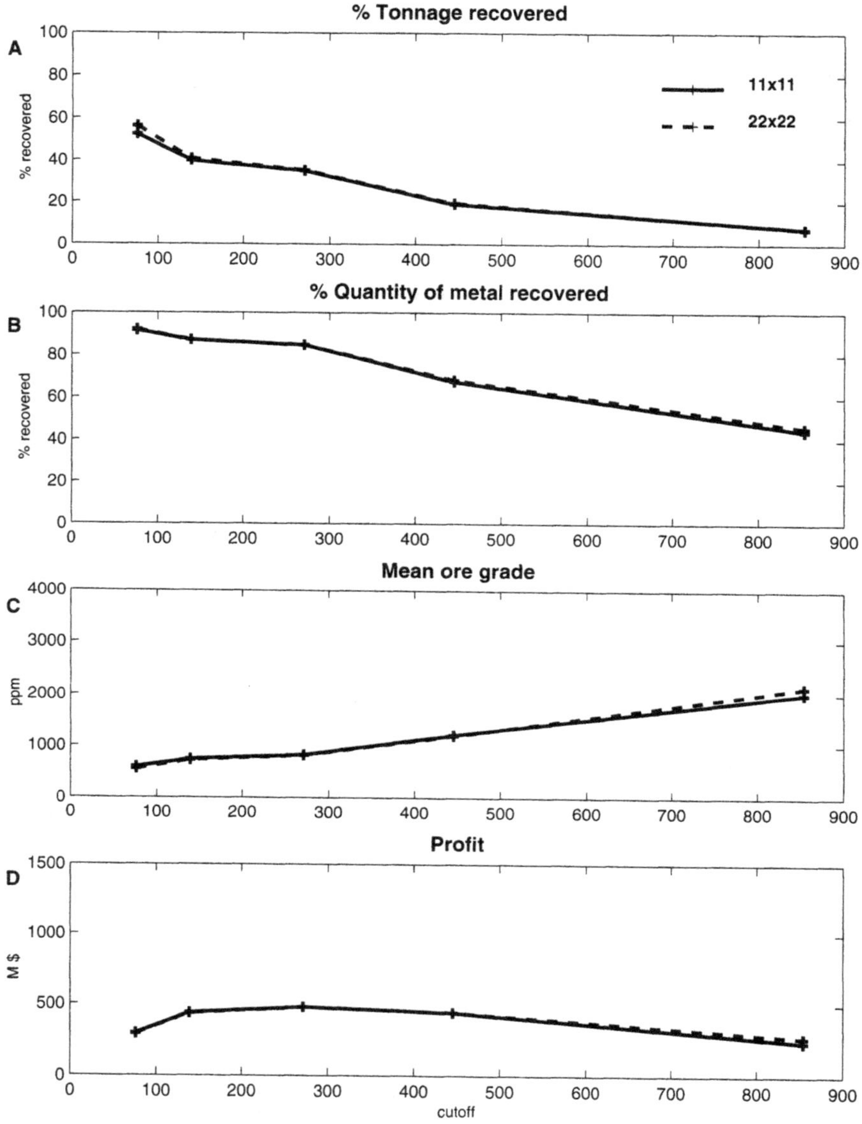

Figure VI.11: Recovery functions (Ely2 data set) for SMU sizes 11×11 and 22×22, estimated via indicator kriging and VVC after compositing. **A:** percent tonnage recovered (total tonnage is for zero cutoff), **B:** percent quantity of metal recovered, **C:** mean ore grade, **D:** profit.

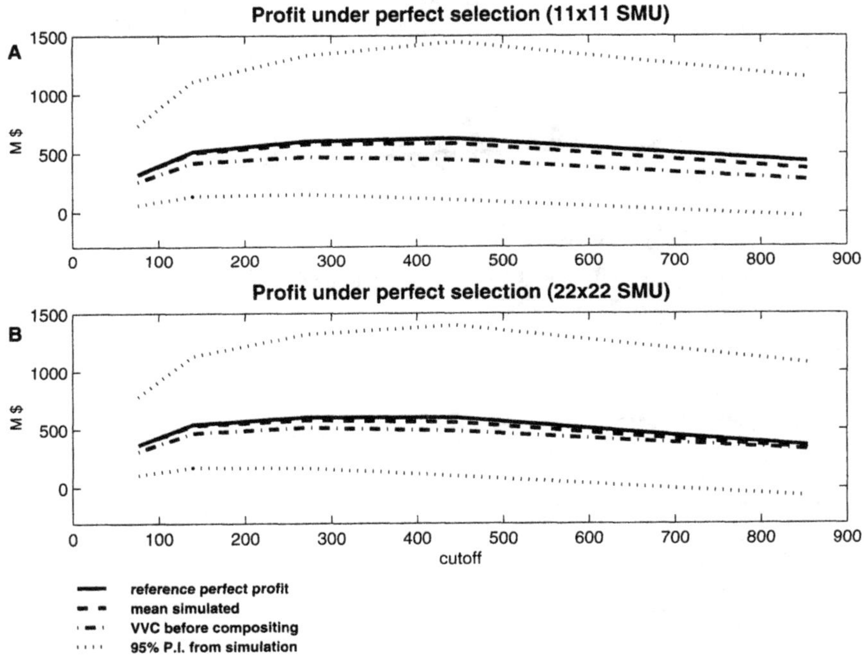

Figure VI.12: Profit under perfect selection (Ely2 data set) for SMU size 11×11 (**A**) and 22×22 (**B**). **Solid line:** true profit, **dashed line:** mean simulated profit, **dash-dotted line:** profit predicted using VVC before compositing, **dotted lines:** 95% probability intervals from the distribution of simulated profits.

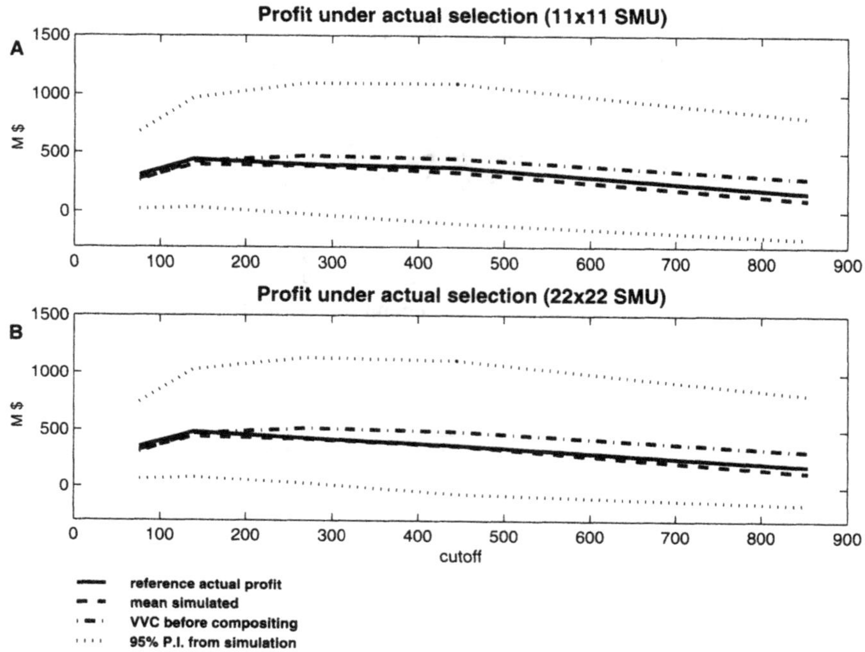

Figure VI.13: Profit under actual selection (Ely2 data set) for SMU size 11×11 (**A**) and 22×22 (**B**). **Solid line:** true profit, **dashed line:** mean simulated profit, **dash-dotted line:** profit predicted using VVC before compositing (hence assuming perfect selection), **dotted lines:** 95% probability intervals from the distribution of simulated profits.

VI.6 Summary

The VVC approach for the support effect was adopted here to provide a measure of comparison for the estimates of perfect recovery given by the simulation approach. Recall that the VVC approach does not account for the information effect, and hence is best used with large SMUs to reduce the efect of misclassification; that is, the effect of the difference between actual SMU grades and their estimates at the time of selection.

One could argue, in the face of the very large uncertainty and corresponding wide 95% probability intervals for all recovery functions, that the VVC estimates are not significantly different from the average simulation estimates. In addition, the VVC approach is significantly easier to implement and has proven to be reasonably robust over numerous applications on a wide variety of mineral deposits.

We could agree with such argument if one could ignore the impact of misclassification; that is, of the information effect: SMUs are selected on the basis of their future estimated values, *not* on their true grade values. For precious metal deposits, where the distribution of very high grades is extremely difficult to predict, the effect of misclassification can be serious and cannot be ignored. Because future estimated SMU grades are not available, they must be simulated together with the true SMU grades. Misclassification can only have a negative effect on profit; there is no balancing of errors! Figure IV.18D indicates that, for the easier Ely data set, such a misclassification can reduce profit from 10% to 40% at the optimal cutoff grades. Such a profit reduction might have been factored in through mining dilution and other safety factors. The simulation approach allows a more rigorous determination of such profit reduction factors; the VVC approach does not offer that facility.

Chapter VII

Conclusions

Many factors make the early prediction of recoverable reserves a difficult challenge that cannot be addressed by mere estimation/interpolation algorithms. These are:

- Data sparsity: when evaluating a new mining venture and the corresponding recoverable reserves, the data available are typically limited to a few drillholes and the related geological interpretation; in many cases bulk ore is not accessible.

- Support effect: future discrimination of ore versus waste will be performed on volumes (SMUs) vastly larger than the volume support of the available data. The statistics (histogram) of SMU grades cannot be predicted straightforwardly from the sample data statistics, nor from the distribution of present SMU estimates because of the smoothing effect.

- Information effect: no selection of ore versus waste can be perfect; some misclassification will necessarily occur. That misclassification depends on the future data available at the time of mining; such data are not available at the time of reserves prediction.

- Accessibility: blocks or entire zones of potential ore may not be recoverable due to mining design and accessibility constraints. The mine design itself depends on the delineation of these recoverable reserves.

- Difficult uncertainty: all the previous factors make uncertainty about recoverable reserves difficult to predict, that uncertainty being much beyond mere estimation or kriging variances.

There is the illusion that all these different factors can be addressed by a good estimation algorithm based on both numerical data (the drillholes) and a sound geological interpretation; the illusion that as long as each local estimate z^* is as accurate as possible given the data available, then the set of all such estimates provides a good proxy for the unknown distribution of grades. More as a measure of safety than to account for uncertainty, "dilution factors" are tacked on the resulting reserves and profit estimates, because experience has shown that reserves based on estimated grades are often too optimistic. Rather than understanding the reason for that bias, mine planning practice has traditionally relied on these dilution factors. The compound effect of all dilution factors can be as high as 50%, which alleviates the effect of the original estimates and may give license to a rushed estimation job. Dilution factors, unless based on production data from the same mining operation, are but stopgaps while awaiting for a more rigorous approach to modeling all factors affecting reserves recovery.

Local versus global accuracy

The illusion that a sound estimation algorithm suffices for ore reserves evaluation comes from the lack of understanding of the trade-offs involved when defining the goodness criterion of an estimate. No estimation algorithm, unless trivially based on exhaustive data, can provide an estimate that is "good" for all purposes. Traditional estimation algorithms, including all distance-based weighting algorithms and kriging, aim at local accuracy; that is, at providing an estimate $z^*(\mathbf{u}_i)$ as close as possible to the corresponding true value $z(\mathbf{u}_i)$ irrespective of its relation with any other estimated value $z^*(\mathbf{u}_j)$, $j \neq i$. This would be fine if the estimation was so good that $z^*(\mathbf{u}_i) \simeq z(\mathbf{u}_i)$ and $z^*(\mathbf{u}_j) \simeq z(\mathbf{u}_j)$, in which case the pair of estimated values $\{z^*(\mathbf{u}_i), z^*(\mathbf{u}_j)\}$ does also reproduce the dependence between the true values $\{z(\mathbf{u}_i), z(\mathbf{u}_j)\}$. Unfortunately, the data available at the time of mine planning and reserves prediction are never enough to consider the approximation $z^*(\mathbf{u}) \simeq z(\mathbf{u})$; the correlation between estimates $z^*(\mathbf{u})$ and true values $z(\mathbf{u})$ rarely exceeds 0.3. In the face of such poor (albeit possibly best) local accuracy, one may consider trading that correlation (possibly down to 0.1) for reproduction of patterns of spatial variability as characterized by a model of dependence between the true values $\{z(\mathbf{u}_i), z(\mathbf{u}_j)\}$. In a simulation approach, the simulated values $\{z^{(l)}(\mathbf{u}_i), z^{(l)}(\mathbf{u}_j)\}$ do not aim at maximizing the correlation with the corresponding true values $\{z(\mathbf{u}_i), z(\mathbf{u}_j)\}$ but at approximating the spatial correlation of these true values; that is:

$$\text{Corr}\{z^{(l)}(\mathbf{u}_i), z^{(l)}(\mathbf{u}_j)\} \simeq \text{Corr}\{z(\mathbf{u}_i), z(\mathbf{u}_j)\}$$

or, equivalently, reproducing the variogram model $\gamma(\mathbf{u}_i - \mathbf{u}_j)$. In advanced geostatistics, one aims at reproducing spatial patterns of many more than two values. The problem is then inference of these patterns, a task typically addressed through the use of training images.

Simulation trades poorer local accuracy for better global or structural accuracy as defined by the reproduction of a model of spatial variability, typically a histogram and a variogram. In presence of limited data, we suggest forfeiting any attempt at locating precisely each ore block or SMU. Instead, one should aim at providing a spatial representation of the grades distribution that mimics fairly the spatial patterns of the true grades. Because these simulated patterns may not be locally accurate, simulation should provide many alternative representations or realizations, all consistent with the few local data available. Average grades z_V of large panels or homogeneous mineralization zones could be estimated because the panels V are large and one could capitalize on the averaging of errors. It would be unreasonable, however, to attempt estimation of small SMU grades within such panels from limited data. Within-panel recovery should be approached through simulation of the spatial patterns of grades distribution within each panel. No localization of the within-panel recovery is yet possible, nor is it useful for mine planning.

Data sparsity
Lack of data is a problem whatever the approach adopted. It is less of a problem for simulation, however, as long as the statistics of the spatial distribution of grades can be established. In a simulation approach, data are needed for two purposes:

- delineation of homogeneous mineralization zones, each defined such that its grade distribution could be characterized by a stationary random function model whose statistics, typically a histogram and a variogram, can be inferred; and
- rough localization of large ore zones.

The data required for the first purpose need not all come from drilling; they can be structural and interpretative in nature. If numerical data are required to build a histogram representative of each mineralization zone, the delineation of that zone and the determination of the corresponding variogram may be guided by geological interpretation, possibly borrowing structural information from outcrops or similar formations mined elsewhere. In modern geostatistics, multiple-point statistics, including and beyond the two-point variogram model, are borrowed from training images, which can be conceptual geological drawings reflecting the grade patterns (clusters, anisotropy) deemed to prevail in each mineralization zone (Strebelle, 2002). Such training images could be available or synthesized from already mined-out areas.

The local conditioning of each simulated realization does require local data, but not to the extent needed for the accurate estimation of each SMU grade.

Support effect
A histogram of SMU grades is needed to evaluate the proportion of such SMUs that could be recovered as ore, yet that histogram cannot be built from estimated SMU grades because of the smoothing effect of estimation. The SMU size and geometry may also vary from one mining face to another. The solution proposed in this book is to simulate the grade distribution at the quasi-point volume support of the available composite data. The simulated

point grades can then be averaged into simulated SMU grades that are pooled into a simulated histogram of SMU grades, from which a simulated realization of ore recovery can be retrieved. Sensitivity of ore recoveries to SMU size and geometry can be easily performed; the availability of a common quasi-point support realization ensures consistency of all results. Had estimation (e.g., kriging) been used, consistency would not have been achieved, as the smoothing effect of estimation is support-dependent: For a given data set, the smaller the SMU size the larger the smoothing effect, essentially through curtailing of the histogram upper tail.

Information effect

Possibly the most important contribution of the simulation approach is the assessment of the effect of misclassification on recovery and, consequently, profit. Selective mining calls for smaller SMUs, possibly of varying size, at a greater production cost: not only are smaller faces more expensive to operate but also better-quality data are needed at the time of mining to accurately select the ore with little dilution. Indeed, SMUs are sorted on the basis of their ultimate estimated values, yet it is the corresponding true grades that are sent to the mill and contribute to the profit. Misclassification is an unavoidable and often critical aspect of selective mining; this may partially explain the present trend toward massive bulk mining.

One can simulate the future selection data, for example, blasthole data, together with the SMU grades, both being conditioned to the present available drillhole data. The simulated blasthole data are then combined into simulated future SMU estimates. Availability of the simulated pairs $\{z_v^{(l)}(\mathbf{u}), z_v^{**(l)}(\mathbf{u})\}$, true grade versus selection estimate, allows an assessment of the effect of misclassification (recall the scattergram of Fig. I.3). Again, sensitivity analysis to various aspects of the information effect is easily performed with consistency ensured by the common quasi-point support simulated grade realizations.

Accessibility

There is rarely if ever free selection: the economic worth of a block or SMU does not depend only on its metal content but also on the cost of reaching it and then mining it, that cost being shared with other neighboring blocks. The decision to mine a block as ore or as waste depends on the mine plan, which itself depends on the estimated grades of that block and its neighbors. Estimation of recoverable reserves and mine planning are thus closely related endeavors that call for a difficult optimization solution. This optimization challenge is never addressed in its entirety. Instead, and traditionally, mine plan and design are based on rough, large-scale estimates of grade distribution without accounting for misclassification; fortunately such large-scale estimates are little affected by the smoothing effect if based on sound prior geological zoning. Within the resulting and necessarily suboptimal mine layouts, estimation of recoverable reserves is then performed, traditionally with little regard for either support or information effect except for dilution factors.

Although this book does not address the problem of optimal mine planning, we believe that simulated realizations of the grades spatial distribution would provide the databases necessary for testing and fine-tuning alternative mine plans. There will come a time when mine planning will reach the level of rigor and scientific repeatability of the design of a new aircraft; at that time simulated models of the grades and other rock characteristic distributions will be needed, and again global or structural accuracy of the model will prevail over local accuracy; that is, simulation will prevail over estimation. In the earth sciences, production of the largest petroleum reservoirs is already routinely optimized through simulation of their flow performance evaluated on geostatistically simulated numerical models of facies geometry and related petrophysical properties.

The solution developed in this book is restricted to free selection within an homogeneous panel V; by homogeneous we mean a panel within which the grade distribution can be simulated through a stationary random function model. Typically, that panel V could be a mineralization zone within the layout of the mine/pit. Free selection within any panel size still represents an approximation. One way around that limitation would be to simulate ore accessibility and recovery within the panel; that simulation would be based on simulated, *not* estimated, realizations of the within-panel grade distributions accounting for both support and information effects. Free selection amounts to retain as ore those SMUs above cutoff grade, constrained selection amounts to further limit selection of those SMUs that are economically accessible. Be aware, though, that accessibility being simulated will change from one simulated realization to another. Indeed, accessibility is an uncertain parameter of ore recovery and mine profitability, and that uncertainty should be, and can be, accounted for in a simulation framework.

Uncertainty assessment

Evaluation of future reserves recovery from early development data is an extremely challenging task, yet one that is critical to mine planning, investment, and ultimately profitability; therefore, an assessment of its uncertainty must be provided. This uncertainty is much beyond the estimation variance of any single block, SMU or panel, as could be provided by kriging; there are many reasons for this, but the most prominent are:

- kriging variances are independent of the data values (homoscedastic) and are no different whether a SMU will be selected as ore or rejected as waste;
- kriging variances of adjacent blocks or SMUs cannot be combined because of the dependence of adjacent estimation errors;
- a variance does not suffice to characterize a distribution (histogram), unless an arbitrary Gaussian distribution is assumed. None of the recovery distributions of Figures V.14–V.21 are Gaussian.

There is no alternative to a simulation approach if the critical uncertainty related to reserves recovery is to be evaluated. The paradigm underlying that uncertainty evaluation is to generate alternative data sets, or simulated realizations, on which the process of imperfect selection is simulated. Provided these simulated data sets mimic reasonably those traits of the actual grade distribution that affect reserves recovery, and provided that we can simulate the future selection process and its related misclassification, we can adopt the distribution (histogram) of the simulated reserves values as a model for the uncertainty about the actual ore reserves (recall Fig. I.1).

The previous provisos render the simulation approach an extremely demanding, and equally rewarding, endeavor that befits the critical importance of reserves assessment. First, the simulated grade distributions must reflect the spatial patterns (continuity-anisotropy) of the actual SMU grades within each mineralization zone or rock type, which requires a sound understanding of the prevailing geology and its correct geostatistical modeling. Second, the future process of mining and ore selection should be reasonably understood to be simulated; this includes simulation of the quality and density of the future selection data. Note that for both simulation processes (geology and mining) various scenarios can, and should, be considered: given an early data set, there can be alternative geological interpretations and many possible scenarios for selective mining.

Workflow and recommendations

No two mineral deposits are exactly similar, and no two selective mining operations are exactly the same. The characteristic of most earth sciences extractive operations is their adaptation to the ever-different aspects of nature, whether a mineral deposit, an oil reservoir, a forest, or a pollution site. There is no unique technology and related software for simulating the spatial distribution of mineral grades, nor is there any catalog of established selective mining procedures; hence, any workflow for the evaluation of mineral resources can be only indicative with tools suggested rather than demonstrated. The strength of geostatistics, more generally statistical tools and more specifically stochastic simulations, is their adaptability to the specific data set and problem encountered.

Yet any geostatistical evaluation of mineral reserves would include the following three different phases, each with varying degrees of involvement:

- data analysis and geological zoning,
- stochastic simulation of grades and the mining process, and
- sensitivity analysis.

The following recommendations, necessarily of general nature, relate to these three different phases.

1. Data analysis and geological zoning

No reserves assessment can be better than the data from which it proceeds. These data include not only the sparse drillhole data but also any ancillary data taken from outcrops or open faces in deemed-similar similar formations, and, most important, geological interpretation. Drillhole data only provide disjointed numerical information—it is their geological interpretation that provides a first global image of what the future mine could be. It is thus of paramount importance that the geological analysis preceding any reserves assessment be as thorough as possible, resulting possibly in different interpretations or scenarios, each consistent with the sparse data available and each leading to possibly different reserves estimates.

Geological definition of structural and mineralization zones is critical for the mine design and corresponding reserves recovery; it is a difficult and highly uncertain task because of data sparsity. Yet, too often in practice, a single geological model (mineralization scenario) is retained, which in turn conditions data interpretation, mine design, reserves recovery, and ultimately mine profitability. Because that geological model will evolve and may even change drastically as the mine is opened, we recommend that alternative geological scenarios be considered in the early phase of reserves assessment, see also the JORC code (JORC, 1999). Each scenario may lead to a different geostatistical model of grade distribution, a different simulation study, and ultimately a different set of reserves estimates: Ely and Ely2 could have provided two such different scenarios, had the initial 30 drillhole data been the same. The difficulty is then to attach a prior probability to each different geological scenario (Journel, 1997).

Rarely, if ever, does a mining operation relate to a single homogeneous mineralization, one that can be characterized by a single stationary random function model. The single most critical result of data analysis and geological interpretation is the definition of mineral zones, whether that definition is based on structural grounds, for example, fault zones, or on mineralization types affecting grades and mineability, for example, rock hardness. Geologists, however, should resist their natural tendency to define too many rock types or mineralization zones. Each zone should be such that its grade distribution (statistics) can be reasonably inferred; this includes at a minimum a representative composite grade histogram and a variogram that could be constructed from geological interpretation of anisotropy directions and ranges. A conceptual variogram model built on sound geological interpretation, with appropriate sensitivity analyses, is better than a quasi-pure nugget effect model fitted on a supposedly objective sample variogram calculated on too few data.

Warning: Sound data and correct geological understanding of the deposit leading to correct delineation of mineralization zones are necessary prerequisites to any sound mine planning and reserves evaluation. No geostatistics can correct data bias, whether originating from ill will or bad luck (recall the remark made at the end of Section V.1), or can undo a wrong geological scenario embedded into the geostatistical model. Any mine planner knows the importance of checking, yet again, data accuracy and considering alternative geological interpretations, particularly those that appear off the mark.

2. Stochastic simulation

The simulation of reserves includes two steps: (i) the geostatistical simulation of the spatial distribution of grades within each mining panel V, and (ii) the simulation of selection and recovery within each such panel V.

A mineralization zone or rock type may include several (possibly disjointed) panels, but each panel relates to a single random function model whose realizations can be simulated with existing geostatistical algorithms. Each realization represents a possible numerical representation of the actual grades distribution within that panel V.

Although recovery of multiple metals is common, actual mine selection rarely considers separately different metal grades; instead, different zones might be considered with a different composite metal grade in each zone. It then suffices to simulate that composite grade for the purpose of simulating reserves recovery, thus avoiding the difficulty of a cosimulation or a coregionalization of different metal grades. Cosimulation may still be needed to evaluate the different mill recovery of each metal.

Typically, the within-panel mineralization is "homogeneously heterogeneous" enough (no complex spatial patterns) to lend itself to a Gaussian-type random function model after normal scores transform. The simulation of such mineralization is then straightforward with a sequential Gaussian simulation algorithm; this requires a grade histogram for the normal score forward and backward transformation, and a normal score variogram model. In the more difficult case of two (rarely more) locally imbricated mineralizations (e.g., Ely2 in this book), an indicator simulation may be required to separate the two mineralizations before their respective simulation with a Gaussian-type algorithm. Again, our recommendation is one of simplicity.

The most challenging aspect of stochastic simulation of reserves recovery is the simulation of the information effect; that is, simulation of the future selection data, selection estimates, and corresponding misclassification. This requires formulating prior scenarios of future mine selectivity including future data quality and density. Such scenarios and their related statistics can be borrowed from deemed-similar selective mining operations. When simulating future data quality, beware that data errors are rarely homoscedastic; that is, independent of the true value signal (at least those data errors that affect misclassification the most).

One need not restrict reserves simulation within a panel to free selection, as long as the mining sequence within that panel can be simulated including all factors, other than accessibility, that can affect ore or material recovery. Again, a simulated database (grades spatial distribution) allows a more rigorous inference of the various components of a dilution factor.

3. Sensitivity analysis

Different simulated realizations of the grade spatial distribution within a given panel V represent already a sensitivity analysis of reserves versus that grade distribution. Those realizations correspond, however, to a random function model whose statistics (histogram-variogram), though uncertain, are frozen. Sensitivity analysis of the effect of those statistics on reserves recovery can be performed by changing, say, the variogram model; this was done for the Ely data set. Similarly, the effect of the within-panel histogram on recoverable reserves can be evaluated by considering alternative histograms obtained by bootstrap from the original sample histogram; such a bootstrap was performed for the Ely data set but was not followed by an analysis of the consequential effect on the simulated grade realizations and the simulated reserves recovery.

Beware not to perform sensitivity analysis on each and every unknown parameter of the geostatistical model or option for the mine selectivity. Given the sparse data available at the time of reserves prediction, almost everything is unknown and it would be unreasonable to try to evaluate all uncertainty. Only those aspects of the reserves assessment study deemed most important, yet uncertain, should be subjected to a sensitivity analysis. The particular geological data interpretation and resulting mineralization zoning is often of much greater importance for metal recovery and profitability than the variogram model used within a particular panel; this is no justification, though, for adopting a simplistic pure nugget effect model everywhere. Our advice is to freeze those parameters that are either deemed less important or are reasonably well known, and to perform sensitivity analysis on the most important parameters or model decisions (such as geological zoning). A formal sensitivity analysis resulting in a probabilistic distribution is actually called stochastic simulation. Thus, if geological zoning is indeed a most critical factor on reserves recovery uncertainty, then it should be formally simulated through a large-scale random function model for the distribution of mineralization zones; such a large-scale simulation of geological facies of an oil reservoir is indeed done before the simulation of porosity and permeability within each lithofacies (Haldorsen and Damsleth, 1990).

Connections with decision theory and risk analysis

Some aspects of recoverable reserves evaluation can be addressed from a decision theoretic perspective, see for example Raiffa and Schlaifer (1961) and Pratt et al. (1995) for an introduction to decision theory. In this framework, each decision alternative, for example, whether to send a SMU to the mill or to the waste pile, or whether to acquire additional blastholes, is tagged with a monetary consequence. Each monetary consequence accounts for acquisition and processing costs, as well as associated benefits. Our simple profit function (IV.4) is along that line of thought, but it does not (cannot) address the monetary consequence of each and every decision involved in mine planning.

Although decision theory methods are well founded and frequently used in engineering design (e.g., Ang and Tang, 1990), their routine application often ignores important aspects of spatial variability and uncertainty in the parameters involved in decisions. For example, the blind export of traditional risk analysis tools from nuclear engineering to mining problems would fail to model the critical importance of spatial correlation of mineral grades. This book contributes to a better appreciation of the importance of spatial correlation, support effect, and information effect in risk analysis studies in a mining context.

One of the most important questions addressed in decision theory is that of the value of additional information. For example, one might be interested in evaluating the worth of data taken from future blastholes; paying for more better-quality blasthole data is warranted if their additional cost does not exceed the expected benefits arising from their acquisition. Conditional to the information available at the present time, one could simulate several different scenarios for the future selection data (quality- and quantity-wise), then evaluate how each scenario performs in approaching perfect selection corresponding to exhaustive sampling with perfectly accurate data.

For applications of decision theory methods to evaluate additional data worth in a spatially distributed setting in other scientific disciplines, the reader is referred to the works of Englund and Heravi (1993) for a soil remediation example, James and Freeze (1993) for a hydrogeological design example, and Solow (1994) for a flood control example.

Uncertainty evaluation, a final quest

One of the initial reasons for proposing the simulation approach and committing to this book was to assess the uncertainty about prediction of mineral reserves. Yet in the "sensitivity analysis" section of this concluding chapter, we recognize that uncertainty is everywhere and that it would be futile to try to evaluate it all. Then, how objective is the final reported measure of uncertainty if some aspects of it have been ignored because they were deemed unimportant, other aspects have been hastily addressed through sensitivity analysis, and only very few aspects have been fully addressed through stochastic simulation? What is the significance of the final probability intervals provided?

Before attempting to provide an answer, we would first like to stress that there is no objective measure of uncertainty. All uncertainty measures are model-based, and that model, which includes decisions about all frozen parameters, is necessarily subjective.

Our opinion is that the exact interval associated with a comprehensive probability (90% or 95%, not 50%) is of no great significance; it suffices to recognize it as small (one could "live" with it) or large (a warning is lit!). In the latter case, it is worth plotting out and having a careful look at the extreme simulated realizations; for example, those yielding the lowest profit or greatest loss. If all realizations are equiprobable and all reproduce the same sample data, what caused those specific realizations to perform so poorly? Is there a possible preventive measure, such as planning a safety stockpile to alleviate simulated breaks in the flow of ore? Or do these extreme realizations appear geologically unreasonable, in which case the random function model underlying the stochastic simulation should be reconsidered or better conditioned to, precisely, prior geological information? Now shouldn't that prior geological information be questioned in the face of those unsettling realizations?

Perhaps the most useful contribution of a conditional simulation exercise, as that proposed in this book, is to force the mine planners to model, and thus recognize and analyze, the specific causes affecting their mining dilution factor values. Such a practice would prevent adopting a somewhat arbitrary dilution factor value, which alone could launch or cancel a mining project. The kind of precise (numerical) modeling called by stochastic simulation would not stand any rushed job brushing away possibly important aspects of uncertainty. At the very least, such a modeling framework requires that all hypotheses about frozen parameters be clearly stated.

The problems addressed in this book are limited to the impact of the support (selection volume) and the information (selection data) effects on mining dilution and reserves evaluation. The scope of the toolkit proposed, stochastic simulation, is much broader: one can simulate many other aspects of the future mining operation, predict average behavior, and hopefully, extreme behavior which may call for preventive measures. Now that computer hardware and (increasingly) software is not anymore an issue, our hope is that the mining industry will catch up with other engineering fields and capitalize on the general concept of virtual simulation of any important mining plan decision, whether at the early stage of a mine planning and reserves evaluation, or at the later stage of grade control.

Appendix A

List of Acronyms

BH: blast holes

ccdf: conditional cumulative distribution function

cdf: cumulative distribution function

DDH: diamond drill holes

IK: indicator kriging

LS: least squares

Median IK: indicator kriging with the same indicator correlogram for all cutoff values

mp: multiple-point, as opposed to two-point (statistics)

RF: random function

RV: random variable

SMU: selective mining unit

USGS: United States Geological Survey

VVC: volume-variance correction

Appendix B

Common Notation

$\forall$: whatever

D: deposit, i.e., an assemblage of large panels V not necessarily all of the same size

$d\mathbf{u}$: an infinitesimal area (volume) centered at location $\mathbf{u}$

$D^2(\cdot/V)$ "dispersion variance" of point support grades within the deposit D

$D^2(v/V)$ "dispersion variance" of grades averaged over selective mining units of volume $|v|$ within the deposit D

V: panel, i.e., an assemblage of N SMUs of equal size (that SMU size may vary from one panel to another); $|V|$ denotes the volume of panel V

$B(\mathbf{u})$: random variable modeling the datum value available at the time of mining (e.g., blasthole grade) at location $\mathbf{u}$

$b(\mathbf{u'}_\alpha)$: future datum value at location $\mathbf{u'}_\alpha$

$E\{\cdot\}$: expected value

$E\{Z(\mathbf{u})|(n)\}$: conditional expectation of the random variable $Z(\mathbf{u})$ given the realizations of n other neighboring random variables (called data)

$\text{Exp}(\mathbf{h})$: exponential semivariogram model, a function of the separation vector $\mathbf{h}$

$F(z)$: stationary cumulative distribution function of a random variable $Z(\mathbf{u})$

$F^{-1}(p)$: inverse cumulative distribution function or quantile function for the probability value $p \in [0, 1]$

$F(\mathbf{u}; z)$: nonstationary cumulative distribution function of the point support random variable $Z(\mathbf{u})$

$F(\mathbf{u}; z|(n))$: nonstationary conditional cumulative distribution function of the point support random variable $Z(\mathbf{u})$, conditioned by the realizations of n other neighboring random variables (called data)

$F_v(\mathbf{u}; z|(n))$: nonstationary conditional cumulative distribution function of the random variable $Z_v(\mathbf{u})$ defined on a selective mining unit support, conditioned by the realizations of n other neighboring random variables (called data)

$\text{Gauss}(\mathbf{h})$: gaussian semivariogram model, a function of the separation vector $\mathbf{h}$

$\gamma(\mathbf{h})$: stationary semivariogram between any two random variables $Z(\mathbf{u})$ and $Z(\mathbf{u}+\mathbf{h})$ separated by vector $\mathbf{h}$

$\gamma_I(\mathbf{h}; z)$: stationary indicator semivariogram between any two indicator random variables $I(\mathbf{u}; z)$ and $I(\mathbf{u}+\mathbf{h}; z)$ defined at cutoff z and separated by vector $\mathbf{h}$

$\gamma_v(\mathbf{h})$: stationary semivariogram between any two random variables $Z_v(\mathbf{u})$ and $Z_v(\mathbf{u}+\mathbf{h})$ averaged over two selective mining units separated by vector $\mathbf{h}$

$\bar{\gamma}(v, v)$: average semivariogram value between any two points discretizing the support v of a selective mining unit

$\bar{\gamma}(V, V)$: average semivariogram value between any two points discretizing the support V of an entire panel

$\mathbf{h}$: separation vector

$I(\mathbf{u}; z)$: binary indicator random variable at location $\mathbf{u}$ for cutoff z

$i(\mathbf{u}; z)$: binary indicator value at location $\mathbf{u}$ for cutoff z

$i_v^0(\mathbf{u}; z_c)$: indicator of perfect selection for a selective mining unit $v(\mathbf{u})$ centered at $\mathbf{u}$

$i_v^{**}(\mathbf{u}; z_c)$: indicator of actual selection for a selective mining unit $v(\mathbf{u})$ centered at $\mathbf{u}$

$\lambda_\alpha(\mathbf{u})$: kriging weight associated with the α-th datum for estimation at location $\mathbf{u}$

m: stationary mean of the random function $Z(\mathbf{u})$

$m(\mathbf{u})$: mean at location $\mathbf{u}$, or non-stationary expected value of the random variable $Z(\mathbf{u})$

$\mu(\mathbf{u}; z)$: Lagrange parameter for indicator kriging at location $\mathbf{u}$ and cutoff z

$m_v^0(z_c)$: mean ore grade at cutoff z_c recovered under perfect selection from selective mining units of volume $|v|$

$m_v(z_c)$: mean ore grade at cutoff z_c recovered under actual selection from selective mining units of volume $|v|$

n: total number of presently available sample data for prediction of recoverable reserves. (n) denotes the set of such samples

$n(\mathbf{u})$: number of presently available sample data within a neighborhood around location $\mathbf{u}$; note that $n(\mathbf{u}) < n$

nn: number of sample data available at the future time of actual selection; note that $nn(\mathbf{u}) >> n(\mathbf{u})$

$N(m(\mathbf{u}), s^2(\mathbf{u}))$: cumulative distribution function of a normal (Gaussian) random variable $Z(\mathbf{u})$ with mean $m(\mathbf{u})$ and variance $s^2(\mathbf{u})$

$p_v^0(z_c)$: profit at cutoff z_c under perfect selection from selective mining units of volume $|v|$

$p_v(z_c)$: profit at cutoff z_c under actual selection from selective mining units of volume $|v|$

$q(p) = F^{-1}(p)$: quantile function or inverse cumulative distribution

$q_v^0(z_c)$: quantity of metal at cutoff z_c recovered under perfect selection from selective mining units of volume $|v|$

$q_v(z_c)$: quantity of metal at cutoff z_c recovered under actual selection from selective mining units of volume $|v|$

$R(\mathbf{u})$: random variable modeling the uncertainty about the discrepancy (error) between the future sample value $b(\mathbf{u})$ and the true grade $z(\mathbf{u})$ at location $\mathbf{u}$

$\rho(\mathbf{h})$: stationary correlogram function

$\rho_I(\mathbf{h}; z_k)$: stationary indicator correlogram function

$\sum_{i=1}^{n} z_i = z_1 + z_2 + \cdots + z_n$: summation of n values

σ^2: point support grade variance

σ_v^2: variance of grades averaged over selective mining units

$t_v^0(z_c)$: ore tonnage at cutoff z_c recovered under perfect selection from selective mining units of volume $|v|$

$t_v(z_c)$: ore tonnage at cutoff z_c recovered under actual selection from selective mining units of volume $|v|$

$\mathbf{u} = (x, y)$: coordinate vector denoting a 2D location within the deposit D. In 3D, $\mathbf{u} = (x, y, e)$, where e denotes the vertical elevation

v: selective mining unit (SMU) within any single panel V; $|v|$ denotes the volume of the SMU v

$Var\{\cdot\}$: variance

$Var\{Z(\mathbf{u})|(n)\}$: conditional variance of the random variable $Z(\mathbf{u})$ given the realizations of n other neighboring random variables (called data)

$y(\mathbf{u})$: normal score transform of a value $z(\mathbf{u})$

z_c, or simply z: cutoff grade above which a SMU is declared as ore, i.e., sent to the mill

$z(\mathbf{u})$: point support grade at location $\mathbf{u}$; the core or composite support volume is assumed constant and quasi-point.

$z(\mathbf{u}_\alpha)$: good quality grade value at sample location $\mathbf{u}_\alpha$ possibly obtained from drill hole (core) data

$z_v(\mathbf{u})$ average grade of the SMU v centered at $\mathbf{u}$, with n_v being the number of points $\mathbf{u}_j$ discretizing the support v

$z_V(\mathbf{u})$ true average grade of the panel V comprised of N SMUs of size $|v|$

$z_v^*(\mathbf{u})$: estimated SMU grade using the $n(\mathbf{u})$ data available at the time of prediction

$z_v^{**}(\mathbf{u})$: future estimated SMU grade using the $nn(\mathbf{u})$ data available at the time of selection

$z^{(l)}(\mathbf{u})$: simulated quasi-point support grade value at location $\mathbf{u}$. The superscript (l) stands for the lth simulated realization out of a total of L: $l = 1, \ldots, L$

$z_v^{(l)}(\mathbf{u})$: simulated true average grade of the SMU v centered at $\mathbf{u}$

$z_v^{**(l)}(\mathbf{u})$: simulated estimated SMU grade using the $nn(\mathbf{u})$ simulated data $\{z^{(l)}(\mathbf{u}'_\alpha),\ \alpha = 1, \ldots, nn(\mathbf{u})\}$, mimicking the information available at the time of selection

Bibliography

Ang, A. S. and W. H. Tang (1990). *Probability Concepts in Engineering Planning and Design. Volume II: Decision, Risk, and Reliability.* John Wiley & Sons, New York.

Agricola, G. (1950). *De Re Metallica* (translation of 1556 original by H. C. Hoover and L. H. Hoover). Dover Publications, New York.

Baafi, E. Y. and N. A. Schofield, editors (1997). *Geostatistics Wollongong '96*, volumes 1 and 2. Kluwer, Dordrecht, The Netherlands.

Barnes, T. E. (1999), Practical post-processing of indicator distributions. In K. Dagdelen, editor, *Proceedings of the 28th International APCOM Symposium*, pages 227–237. Society of Mining Engineers, Golden, CO.

Bideaux, R. (1990). Ore reserve estimation. In *Surface Mining*, 2nd edition, pages 287–384. Society for Mining, Metalurgy, and Exploration, Littleton, CO.

Caers, J., J. Gelders, A. Vervoot, and L. Rombouts (1997). Non-conditional and conditional simulation of a spatial point process: Application to diamond deposits. In E. Y. Baafi and N. A. Schofield, editors, *Geostatistics Wollongong '96*, volume 1, pages 270–281. Kluwer, Dordrecht, The Netherlands.

Caers, J. K. and A. G. Journel (1998). Stochastic reservoir simulation using neural networks trained on outcrop data. SPE paper no. 49026.

Caers, J. and M. Maes (1998). Statistics for deciding on tail behaviour, bounds and end-points. *Structural Safety*, 20:1–23.

Chilès, J. P. (1984). Simulation of a nickel deposit: Problems encountered and practical solutions. In G. Verly, M. David, A. G. Journel, and A. Maréchal, editors, *Geostatistics for Natural Resources Characterization*, Part 2, pages 1015–1030. Reidel, Dordrecht, Holland.

Chilès, J. P. and P. Delfiner (1999). *Geostatistics: Modeling Spatial Uncertainty.* John Wiley & Sons, New York.

Dagbert, M. (1980). The use of simulated spatially distributed data in geology. *Computers and Geosciences*, 6(2):175–192.

Damsleth, E., C. B. Tjølsen, H. Omre, and H. H. Haldorsen (1992). A two-stage stochastic model applied to a North Sea reservoir. *Journal of Petroleum Technology*, April 1992:402-408.

David, M. (1972). Grade-tonnage curve: Use and misuse in ore reserve calculation. *Transactions of the Institute of Mining and Metallurgy, Sect. A: Min. Industry*, A129–A132.

David, M. (1977). *Geostatistical Ore Reserve Estimation*. Elsevier, Amsterdam.

David, M. (1988). *Handbook of Applied Advanced Geostatistical Ore Reserve Estimation*. Elsevier, Amsterdam.

Davis, B. M. (1984). Indicator kriging as applied to an alluvial gold deposit. In G. Verly, M. David, A. G. Journel, and A. Maréchal, editors, *Geostatistics for Natural Resources Characterization*, Part 1, pages 337–348. Reidel, Dordrecht, Holland.

Davis, J. C. (1973). *Statistics and Data Analysis in Geology*. John Wiley & Sons, New York.

Deraisme, J., and R. Dumay (1979). Geostatistics and mining processes. In T. J. O'Neil, editor, *Proceedings of the 16th International APCOM Symposium*, pages 149–162. Society of Mining Engineers, New York.

Deraisme, J., C. de Fouquet, and H. Fraisse (1984). Geostatistical ore body model for computer optimization of profits from different underground mining methods. In *Proceedings of the 18th International APCOM Symposium*, pages 583–590. Institute of Mining and Metallurgy, London.

Deutsch, C. V. (2002). *Geostatistical Reservoir Modeling*. Oxford University Press, New York.

Deutsch, C. V. and A.G. Journel (1998). *GSLIB: Geostatistical Software Library and User's Guide*, 2nd edition. Oxford University Press, New York.

Deutsch, C. V., E. V. Magri, and K. P. Norrena (2000). Optimal grade control using geostatistics and economics: Methodology and examples. *Society of Mining Engineers SME Transactions*, (308):43–52.

Dimitrakopoulos, R. (1998). Conditional simulation algorithms for modelling orebody uncertainty in open pit optimization. *International Journal of Surface Mining*, 12(4):173–179.

Dimitrakopoulos, R. and M. Dagbert (1993). Sequential modeling of relative indicator variables: Dealing with multiple lithologic types. In A. Soares, editor, *Geostatistics Tróia '92*, volume 1, pages 413–424. Kluwer, Dordrecht, The Netherlands.

Dowd, P. (1994). Risk assessment in estimation and open pit planning. *Transactions of the Institute of Mining and Metallurgy, Sect. A: Min. Industry*, 103:A148–A154.

Dowd, P. (1997). Structural controls in the geostatistical simulation of mineral deposits. In E. Y. Baafi and N. A. Schofield, editors, *Geostatistics Wollongong '96*, volume 1, pages 647–657. Kluwer, Dordrecht, The Netherlands.

Efron, B. (1982). *The Jackknife, the Bootstrap, and other Resampling Plans.* Society for Industrial and Applied Mathematics, Philadelphia, PA.

Efron, B. and R. J. Tibshirani (1993). *An Introduction to the Bootstrap.* Chapman & Hall, New York.

Englund, E. J. and N. Heravi (1993). Conditional simulation: Practical application for sampling design optimization. In A. Soares, editor, Geostatistics Tróia '92, volume 2, pages 613–624. Kluwer, Dordrecht, The Netherlands.

Farrelly, C. and R. Dimitrakopoulos (2002). Recoverable reserves and support effects in optimizing open pit mine design. *International Journal of Surface Mining*, 16(3):217–229.

Glacken, I. M. (1997). Change of support by direct conditional block simulation. In E. Y. Baafi and N. A. Schofield, editors, *Geostatistics Wollongong '96*, volume 1, pages 811–821. Kluwer, Dordrecht, The Netherlands.

Goméz-Hernández, J., A. Soares, and R. Froidevaux, editors (1999). *geoENV II: Geostatistics for Environmental Applications.* Kluwer, Dordrecht, The Netherlands.

Goovaerts, P. (1997). *Geostatistics for Natural Resources Evaluation.* Oxford University Press, New York.

Guardiano, F. and R. M. Srivastava (1993). Multivariate geostatistics: Beyond bivariate moments. In A. Soares, editor, *Geostatistics Tróia '92*, volume 1, pages 133–144. Kluwer, Dordrecht, The Netherlands.

Guardiano, F. B., H. M. Parker, and E. H. Isaaks (1995). Prediction of recoverable reserves using conditional simulation: A case study for the Fort Knox gold project, Alaska. Unpublished report, Mineral Resources Development Inc., Foster City, CA.

Gy, P. (1998). *Sampling for Analytical Purposes.* John Wiley & Sons, London.

Haldorsen, H. and E. Damsleth (1990). Stochastic modeling. *Journal of Petroleum Technology*, April 1990:404–412.

Isaaks, E. H. and R. M. Srivastava (1989). *An Introduction to Applied Geostatistics*. Oxford University Press, New York.

James, B. R. and R. A. Freeze (1993). The worth of data in predicting aquitard continuity in hydrogeological design. *Water Resources Research*, 29(7):2049–2065.

JORC (Joint Ore Reserves Committee) (1999). *Australasian code for reporting of mineral resources and ore reserves*. The Australasian Institute of Mining and Metallurgy, Melbourne, Australia.

Journel, A. G. (1974). Geostatistics for conditional simulation of ore bodies. *Economic Geology*, 69(5):673–687.

Journel, A. G. (1983). Non-parametric estimation of spatial distributions. *Mathematical Geology*, 15(3):445–468.

Journel, A. G. (1985). The deterministic side of geostatistics. *Mathematical Geology*, 17(1):1–15.

Journel, A. G. (1989). *Fundamentals of Geostatistics in Five Lessons*, Volume 8 Short Course in Geology. American Geophysical Union, Washington, D.C.

Journel, A. G. (1993). Geostatistics: Roadblocks and challenges, In A. Soares, editor, *Geostatistics Tróia '92*, volume 1, pages 213–224. Kluwer, Dordrecht, The Netherlands.

Journel, A. G., (1997). The abuse of principles in model building and the quest for objectivity. In E. Y. Baafi and N. A. Schofield, editors, *Geostatistics Wollongong '96*, volume 1, pages 3–14. Kluwer, Dordrecht, The Netherlands.

Journel, A. G. and C.J. Huijbregts (1978). *Mining Geostatistics*. Academic Press, New York.

Journel, A. G. and E. H. Isaaks (1984). Conditional indicator simulation: Application to a Saskatchewan uranium deposit. *Mathematical Geology*, 16(7):685–718.

Journel, A. G. and F. G. Alabert (1989). Non-Gaussian data expansion in the earth sciences, *Terra Nova*, 1:123–134.

Journel, A. G. and C. V. Deutsch (1993). Entropy and spatial disorder. *Mathematical Geology*, 25(3):329–355.

Journel, A. G., P. C. Kyriakidis, and S. Mao (2000). Correcting the smoothing effect of estimators: A spectral post-processor. *Mathematical Geology*, 32(7):787–813.

Kennedy, B. A., editor (1990). *Surface Mining*, 2nd edition. Society for Mining, Metalurgy, and Exploration, Littleton, CO.

Kim, Y. C., P. K. Medhi, and A. Arik (1982). Investigation of in-pit ore-waste selection procedures using conditionally simulated orebodies. In T. B. Johnson, and R. J. Barnes, editors, *Proceedings of the 17th International APCOM Symposium*, pages 121–142. Society of Mining Engineers, New York.

King, H. F., D. W. McMahon, and G. J. Bujtor (1982). A guide to understanding ore reserve estimation. Australian Institute of Mining and Metallurgy. Supplement to proceedings No. 281.

Koch, G. S. and R. F. Link (1970-71). *Statistical Analysis of Geological Data*, volumes 1 and 2. John Wiley & Sons, New York.

Krige, D. G. (1951). A statistical approach to some mine valuations and allied problems at the Witwatersrand. Master's thesis, University of Witwatersrand, South Africa.

Kyriakidis, P. C. (1997). Selecting panels for remediation in contaminated soils via stochastic imaging. In E. Y. Baafi and N. A. Schofield, editors, *Geostatistics Wollongong '96*, volume 2, pages 973–983. Kluwer, Dordrecht, The Netherlands.

Lane, K. F. (1988). *The Economic Definition of Ore: Cut-off Grades in Theory and Practice.* Mining Journal Books, London.

Lerchs, H. and I. F. Grossmann (1965). Optimum design of open pit mines. *C.I.M. Bulletin*, 58(633):47–54.

Matheron, G. (1962). Traité de géostatistique appliquée. volume 1 (1962), volume 2 (1963), ed. Technip, Paris.

Matheron, G. (1984). The selectivity of distributions and the "second principle of geostatistics". In G. Verly, M. David, A. G. Journel, and A. Maréchal, editors, *Geostatistics for Natural Resources Characterization*, part 1, pages 421–433. Reidel, Dordrecht, Holland.

Norrena, K. P. and C. V. Deutsch (2001). Determination of dig limits subject to geostatistical, economic and equipment constraints. In *Society of Mining Engineers SME Meeting.* February 2001, Denver, CO.

Nowak, M. S., R. M. Srivastava, and A. J. Sinclair (1993). Conditional simulation: A mine planning tool for a small gold deposit. In A. Soares, editor, *Geostatistics Tróia '92*, volume 2, pages 977–988. Kluwer, Dordrecht, The Netherlands.

Nowak, M. S. and R. M. Srivastava (1997). A geostatistical conditional simulation algorithm that exactly honors a predefined grade-tonnage curve. In E. Y. Baafi and N. A. Schofield, editors, *Geostatistics Wollongong '96*, volume 2, pages 669–682. Kluwer, Dordrecht, The Netherlands.

Parker, H. M. (1980). The volume-variance relationship: A useful tool for mine planning. In P. Mousset-Jones, editor, *Geostatistics*, pages 61–91. McGraw Hill, New York.

Parker, H. M. (1991). Statistical treatment of outlier data in epithermal gold deposit reserve estimation. *Mathematical Geology*, 23(2):175–200.

Parker, H. M. and Switzer, P. (1975). The use of conditional probability distributions in ore reserve estimation: A case study. In *Proceedings of the 13th International APCOM Symposium*, pages MII–1–MII–16. Society of Mining Engineers, Clausthal, Germany.

Pitard, F. (1993). *Pierre Gy's Sampling Theory and Sampling Practice*, 2nd edition. CRC Press, Boca Raton, FL.

Pratt, J. W., H. Raiffa, and R. Schlaifer (1995). *Introduction to Statistical Decision Theory*. MIT Press, Cambridge, MA.

Raiffa, H. and R. Schlaifer (1961). *Applied Statistical Decision Theory*. Harvard Business School, Boston, MA.

Ravenscroft, P. J. (1992). Risk for mine scheduling by conditional simulation. *Transactions of the Institute of Mining and Metallurgy, Sect. A: Min. Industry*, 101:A104–A108.

Ravenscroft, P. J. (1993). Conditional simulation for mining. In M. Armstrong and P. Dowd, editors, *Geostatistical Simulations*, pages 79–87. Kluwer, Dordrecht, The Netherlands.

Rendu, J.-M. (1979). Normal and lognormal estimation. *Mathematical Geology*, 11(4):407–422.

Rendu, J.-M and L. A. Ready (1982). Geology and the semivariogram: A critical relationship. In T. B. Johnson, and R. J. Barnes, editors, *Proceedings of the 17th International APCOM Symposium*, pages 771–783. Society of Mining Engineers, New York.

Rossi, M. and H. Parker (1994). Is there light at the end of the tunnel in recoverable reserves?. In R. Dimitrakopoulos, editor, *Geostatistics for the Next Century*, pages 259–276. Kluwer, Dordrecht, The Netherlands.

Rossi, M. (1999). Improving the estimates of recoverable reserves. *Mining Engineering*, 51(1):50–54.

Sichel, H. S. (1952). New methods in the statistical evaluation of mine sampling data. *Transactions of the Institute of Mining and Metallurgy*, 61:261–288.

Solow, A. R. (1994). Conditional simulation and the value of information. In R. Dimitrakopoulos, editor, *Geostatistics for the Next Century*, pages 209–217. Kluwer, Dordrecht, The Netherlands.

Springett, M. W. (1988). Biasing unbiased data: Mine versus mill. In *Proceedings of the 21st International APCOM Symposium*, pages 287-299. Society of Mining Engineers, New York.

Srivastava, R. M. (1987). Minimum variance or maximum profitability? *CIM Bulletin*, 80(901):63–68.

Srivastava, R. M. (1994a). An annealing procedure for honoring change of support statistics. In R. Dimitrakopoulos, editor, *Geostatistics for the Next Century*, pages 277–290. Kluwer, Dordrecht, The Netherlands.

Srivastava, R. M. (1994b). The visualization of spatial uncertainty. In J. M. Yarus and R. L. Chambers, editors, *Stochastic Modeling and Geostatistics: Principles, Methods, and Case Studies*, pages 339–345. American Association of Petroleum Geologists, Tulsa, OK.

Strebelle, S. (2002). Conditional simulation of complex geological structures using multiple-point statistics. *Mathematical Geology*, 34(1):1–21.

Whittle, J. (1997). *Optimization in Mine Design*. WH Bryan Mining Geology Research Center, Brisbane, Australia.

Yarus, J. M. and R. L. Chambers, editors (1995). *Stochastic Modeling and Geostatistics: Principles, Methods, and Case Studies*. American Association of Petroleum Geologists, Tulsa, OK.

Index

www.ingramcontent.com/pod-product-compliance
Lightning Source LLC
LaVergne TN
LVHW011801200926
848484LV00003B/748